ISBN 978-1-716-46500-0
Imprint: Lulu.com

Basic procedures in ordinary differential equations

February 23, 2021

Contents

Preface

The main idea of this book is for the students to learn and practice some basic procedures used to understand ordinary differential equations. Since the main goal for the book is for the reader to get familiar with the material, the book has very little applications. Proofs are provided only when they are easy to follow and contribute to the understanding of the method being explained at the moment.

The author is aware that some of the examples solved in the textbook are somewhat repetitive. This is done with the intension that the reader practice the algebra and techniques that are required to solve the problems. The author trusts that the reader will decide not to go over examples or homework problems once he/she realizes that they are essentially the same as a previous problem.

chapter 1

Definition of differential equation

This chapter goes over some background concepts that are used over and over during the techniques developed in this book. This chapter also provides the definition of differential equations and first examples.

1.1 Basic notions

In this section we go over the notion of real and complex numbers and some of their properties. We also review solutions of quadratic equations because they will be fundamental in several procedures developed in this book. This section also reviews basic derivatives and antiderivatives.

1.1.1 Real numbers

The set of *real numbers* is denoted by $\mathbb{R}$ and every element in $\mathbb{R}$ can be viewed as:

a positive or negative sign, followed by a sequence of finitely many digits, followed by a dot, followed by a sequence of infinitely many decimals.

The sequence of digits after the dot is called the *decimal part* of a number. For example the number -2 can be viewed as $-2.00\ldots$, the number 5.23 can be view as $+5.2300\ldots$. Usually when the sign in front of the number is $+$, this sign is omitted. With this point of view all the numbers are not much different since they require the same effort to write. For example $\pi = 3.14159265\ldots$ and $0 = 0.00\ldots$ and $e = 2.718281828459045235\ldots$ and $\frac{1}{3} = 0.33\ldots$. We want to point out that there may be two ways to write the same number. For example $5 = 5.0\ldots$ or $5 = 4.99\ldots$. A number n with decimal part equal to either $00\ldots$ or $99\ldots$ is called an integer or a whole number. Many times numbers are defined by a property that they satisfy. For example, the only positive solution of the equation $x^2 = 2$ is denoted by $\sqrt{2}$, and we may say that nobody knows the decimal part of $\sqrt{2}$ but there are mathematical methods to compute as many decimal digits of $\sqrt{2}$ as we want. We have $\sqrt{2} = 1.414213562373095048\ldots$. The number $\ln(3)$ is the only solution of the equation $e^x = 3$, we have $\ln(3) = 1.098612288668109691\ldots$.

1.1.2 Properties of real numbers

Here are some basic properties of exponents.

Proposition 1 *Let a and b be positive real numbers, we have*

1. $a^{n+m} = a^n a^m$, *for any real numbers n and m.*

2. $a^{-n} = \frac{1}{a^n}$, *for any real numbers n.*

3. $(a^n)^m = a^{nm}$, *for any real numbers n and m.*

4. $(ab)^n = a^n b^n$, *for any real numbers n.*

5. $\sqrt[m]{a^n} = a^{\frac{n}{m}}$, *for any real numbers n and any positive integer m.*

6. $e^{n \ln(b)} = b^n$, *for any real numbers n.*

1.1.3 Derivative of a function

For any real-valued function $f(t)$ with domain on an interval that contains t_0 we define

$$f'(t_0) = \frac{df}{dt}(t_0) = \lim_{h \to 0} \frac{f(t_0 + h) - f(t_0)}{h}$$

provided the limit exists. As an example, if $f(t) = t^2$ then, for any t_0 we have that

$$\frac{f(t_0 + h) - f(t_0)}{h} = \frac{(t_0 + h)^2 - t_0^2}{h} = \frac{2t_0 h + h^2}{h} = 2t_0 + h$$

provided $h \neq 0$. Therefore,

$$f'(t_0) = \lim_{h \to 0} 2t_0 + h = 2t_0$$

Most of the time we avoid using the new variable t_0 and we just write $\frac{dt^2}{dt} = 2t$. It is clear that when the derivative of a function $f(t)$ exists for every $t \in I$, with I an interval, then $f'(t)$ is a function on I. If the derivative of $f'(t)$ exists, then this derivative is denote by $f''(t)$ or $\frac{d^2 f}{dt^2}(t)$. If the derivative of $f''(t)$ exists, then this derivative is denote by $f'''(t)$ or $\frac{d^3 f}{dt^3}(t)$. In the same way we define $\frac{d^n f}{dt^n}(t)$ for any positive integer n. If f is a function whose derivative $\frac{d^n f}{dt^n}(t)$ exists for all positive integer n then, this function is called a *smooth function*.

Usually we do not compute derivatives by computing the limit provided by the definition, but by applying properties.

> **Proposition 2** *1. If $f(t) = c$ is a constant function then $f'(t) = 0$ for all t.*
>
> *2. If c is a constant and f is a function, then $\frac{dcf}{dt}(t) = c\frac{df(t))}{dt}$*
>
> *3. If $f(t) = t^n$ then $f'(t) = nt^{n-1}$. We also write $\frac{dt^n}{dt} = nt^{n-1}$.*
>
> *4. $\frac{d\sin(t)}{dt} = \cos(t)$, $\frac{d\cos(t)}{dt} = -\sin(t)$, $\frac{de^t}{dt} = e^t$, $\frac{d\ln t}{dt} = \frac{1}{t}$, $\frac{d\arctan(t)}{dt} = \frac{1}{1+t^2}$.*
>
> *5. $\frac{d(f+g)}{dt}(t) = \frac{df}{dt}(t) + \frac{dg}{dt}(t)$.*
>
> *6. $\frac{d(fg)}{dt}(t) = \frac{df}{dt}(t)\,g(t) + \frac{dg}{dt}(t)\,f(t)$. This property is called the product rule.*
>
> *7. $\frac{d(f/g)}{dt}(t) = \frac{\frac{df}{dt}(t)\,g(t) - \frac{dg}{dt}(t)\,f(t)}{g^2(t)}$. This property is called the quotient rule.*
>
> *8. $\frac{d(f(g(t)))}{dt} = \frac{df}{dt}(g(t))\frac{dg}{dt}(t)$. This property is called the chain rule.*

Example 1 *If $f(t) = \cos(3t)e^{2t}$, then $f'(t) = -3\sin(3t)e^{2t} + 2\cos(3t)e^{2t}$ and*

$$f''(t) = -9\cos(3t)e^{2t} - 6\sin(3t)e^{2t} - 6\sin(3t)e^{2t} + 4\cos(3t)e^{2t} = -5\cos(3t)e^{2t} - 12\sin(3t)e^{2t}$$

Example 2 *If A B and C are constants and $f(t) = A + Bt + Ce^{-3t}$, then $f'(t) = B - 3Ce^{-3t}$ and $f''(t) = 9Ce^{-3t}$*

1.1.4 Antiderivatives

The inverse process of doing a derivative is called antiderivative. To be precise we say that $F(t)$ is an antiderivative of $f(t)$ if $F'(t) = f(t)$. Notice that if $F(t)$ is an antiderivative of $f(t)$ and c is a constant, then $F(t) + c$ is also an antiderivative. We will write $\int f(t)\,dt = F(t) + c$ or sometimes we just write $\int f(t)\,dt = F(t)$. In the same way there are some properties that allow us to take derivatives, there are some properties that allow us to compute some antiderivatives

> **Proposition 3**   *1. If $n \neq -1$, then $\int t^n \, dt = \frac{t^{n+1}}{n+1}$.*
>
> 2. *$\int \frac{1}{t} \, dt = \ln(|t|)$*
>
> 3. *$\int \cos(t) \, dt = \sin(t)$, $\int \sin(t) \, dt = -\cos(t)$, $\int e^t \, dt = e^t$, $\int \frac{1}{1+t^2} \, dt = \arctan(t)$*
>
> 4. *$\int f(u(t))u'(t) \, dt = F(u(t))$ where $F(u)$ is an antiderivative of $f(u)$. This property is called integration by substitution and sometimes we use it as follows: Let us assume that we want to do the integral $\int G(t) \, dt$ and we realize that we can find a function $u = u(t)$ such that by replacing $u'(t) \, dt$ with du and $u(t)$ with u we can change the expression $\int G(t) \, dt$ into $\int f(u) \, du$, then an antiderivative of $G(t)$ is $F(u(t))$, provided that $F'(u) = f(u)$.*
>
> 5. *Using the notation $du = u'(t)dt$ and $dv = v'(t)dt$, we have that for any pair of functions $u(t)$ and $v(t)$*
>
> $$\int u \, dv = uv - \int v \, du$$
>
> *This property is called integration by parts.*

Let us see some examples.

Example 3 *If we want to solve $\int \frac{5}{2+3t} \, dt$ then by making the substitution $u = 2 + 3t$ we obtain that $du = 3 \, dt$ and therefore*

$$\int \frac{5}{2+3t} \, dt = \int \frac{5}{u} \frac{du}{3} = \frac{5}{3} \int \frac{du}{u} = \frac{5}{3} \ln(u) = \frac{5}{3} \ln(|2 + 3t|)$$

Notice that if for some reason we know that $2 + 3t$ is positive then we write $\int \frac{5}{2+3t} \, dt = \frac{5}{3} \ln(2 + 3t)$ and if for some reason we know that $2 + 3t$ is negative then we write $\int \frac{5}{2+3t} \, dt = \frac{5}{3} \ln(-2 - 3t)$.

Example 4 *If we want to solve $\int \frac{2}{1+4t^2} \, dt$, it would be wrong to say that this integral is $2 \ln(1+4t^2)$. We can solve the integral $\int \frac{2}{1+4t^2} \, dt$ by making the substitution $u = 2t$, then $du = 2 \, dt$ and therefore*

$$\int \frac{2}{1 + 4t^2} \, dt = \int \frac{du}{1 + u^2} = \arctan(u) = \arctan(2t)$$

Example 5 *For any $a \neq 0$, the integral $\int e^{at} \, dt$ can be solved by making $u = at$. We have that $du = a \, dt$ and therefore*

$$\int e^{at} \, dt = \frac{1}{a} \int e^u \, du = \frac{1}{a} e^u = \frac{1}{a} e^{at}$$

The same substitution we can show that $\int \sin(at) \, dt = -\frac{1}{a} \cos(at)$ and $\int \cos(at) \, dt = \frac{1}{a} \sin(at)$.

Example 6 *The integral $\int te^{2t} \, dt$ can be solved by integration by parts. We can make $u = t$ and $v = e^{2t}/2$, therefore, $du = dt$ and $dv = e^{2t} \, dt$. We have,*

$$\int te^{2t} \, dt = \int u \, dv = uv - \int v \, du = t\frac{e^{2t}}{2} - \int \frac{1}{2}e^{2t} \, dt = t\frac{e^{2t}}{2} - \frac{1}{4}e^{2t}$$

1.1.5 Complex numbers

In this subsection we review the multiplication and addition of complex numbers. We also define e^z when z is a complex number.

Definition 1 *A complex number z is a number of the form $z = a + ib$ with a and b real numbers. The real number a is called the real part of z and it is denoted by $\mathrm{Re}[z]$. The real number b is called the imaginary part of z and it is denoted as $\mathrm{Im}[z]$. We do the sum and product of two complex numbers by treating i as a variable and changing i^2 for -1 whenever it shows up. More precisely we have that*

$$(a + ib) + (c + id) = (a + c) + (b + d)i \quad and \quad (a + ib)(c + di) = ac - bd + i(ad + bc)$$

Example 7 *We have that*

$$(2 + i)(3 - 5i) = 6 - 10i + 3i - 5i^2 = 6 - 7i - 5(-1) = 6 - 7i + 5 = 11 - 7i$$

Example 8 *If t is a real number, we have that*

$$\left(2\sin(t) + ie^t\right)(3t - 5\cos(t)\,i) = 6t\sin(t) - 10\sin(t)\cos(t)\,i + 3te^t\,i + 5e^t\cos(t)$$

Therefore,

$$\mathrm{Re}\!\left[\left(2\sin(t) + ie^t\right)(3t - 5\cos(t)\,i)\right] = 6t\sin(t) + 5e^t\cos(t)$$

and

$$\mathrm{Im}\!\left[\left(2\sin(t) + ie^t\right)(3t - 5\cos(t)\,i)\right] = -10\sin(t)\cos(t) + 3te^t$$

Theorem 1 *If $b^2 - 4c < 0$ then the solution of the equation $\lambda^2 + b\lambda + c = 0$ are the two complex numbers*

$$z_1 = -\frac{b}{2} + \frac{\sqrt{4c - b^2}}{2}\,i \quad and \quad z_2 = -\frac{b}{2} - \frac{\sqrt{4c - b^2}}{2}\,i$$

Example 9 *We have that a solution of the equation $\lambda^2 + 2\lambda + 2 = 0$ is $z = -\frac{2}{2} + \frac{\sqrt{4}}{2}i = -1 + i$. We can directly check this,*

$$z^2 + 2z + 2 = (-1 + i)^2 + 2(-1 + i) + 2 = (1 - 2i + i^2) - 2 + 2i + 2 = 0$$

The other solution of the equation is $z = -1 - i$.

Definition 2 *For any real number θ we define*

$$e^{i\theta} = \cos\theta + i\sin\theta$$

If a, b and t are real numbers, then

$$e^{at + ibt} = e^{at}\left(\cos(bt) + i\sin(bt)\right)$$

Example 10 *Let us compute the real and imaginary part of* $(3 + 2i)e^{4t+5ti}$. *We have*

$$
\begin{aligned}
(3 + 2i)e^{4t+5ti} &= (3 + 2i)(e^{4t}(\cos(5t) + i\sin(5t))) \\
&= e^{4t}(3\cos(5t) - 2\sin(5t) + i(3\sin(5t) + 2\cos(5t)))
\end{aligned}
$$

Therefore

$$
\mathrm{Re}((3 + 2i)e^{4t+5ti}) = e^{4t}(3\cos(5t) - 2\sin(5t)) \quad \text{and}
$$

$$
\mathrm{Im}((3 + 2i)e^{4t+5ti}) = e^{4t}(3\sin(5t) + 2\cos(5t))
$$

1.2 Notion of differential equation

Let us start this section with the definition of differential equation.

Definition 3 *An **Ordinary Differential Equation, an ODE**, is an equation whose solution is a smooth function $y = y(t)$ defined on an interval or union of intervals and, for some $k \geq 1$ the function $\frac{d^k y}{dt^k}$ is present in the equation. If $\frac{d^n y}{dt^n}$ is the highest derivative present in the equation we call the equation an order n ordinary differential equation.*

Example 11 *The equation $y(t)\sin(y(t)) + \cos(y(t)) = 2t^2$ is not a differential equation because for any $k > 0$, $\frac{d^k y}{dt^k}$ is not present in the equation.*

Example 12 *The equation $(t^2 + 1)\frac{dy}{dt} = 5$ is a differential equation of order 1.*

Example 13 *The equation $\frac{dy}{dt} = 5y^2 + 2y$ is a differential equation of order 1.*

Example 14 *The equation $y\frac{d^2 y}{dt^2} = 5t + 2\frac{dy}{dt}$ is a differential equation of order 2.*

Definition 4 *A differential equation where the independent variable of the function that we are looking for (the variable t) is not present, is called an **autonomous equation.***

Notice that Example 13 is an autonomous equation while Example 12 is not autonomous.

Remark 1 *When $y(t)$ represents the position of a particle that moves on a line, then $\frac{dy}{dt}$ represents its velocity. If we can solve for $\frac{dy}{dt}$ and write $\frac{dy}{dt} = f(y, t)$ then the expression $f(y, t)$ is called the velocity field. If the expression $f(y, t)$ does not depend on t then the differential equation is autonomous or in fluid mechanics terminology, the vector field is called a **steady vector field**.*

In general, to solve an equation is not an easy task, even when the equation is not a differential equation but an equation that is searching for a number. For example the equation $x^5 + 2x^3 - x + 1 = 0$ that searches for a real number x is a difficult one. We would like to point out that even though solving an equation may be difficult, checking if a value for the variable is a solution is in general easy. For example we can say that $x = 2$ is not a solution for $x^5 + 2x^3 - x + 1 = 0$ because $2^5 + 2 * 2^3 - 2 + 1 \neq 0$. The same statement holds true for a differential equation. for example, it is easy to check that the function $y(t) = t^2$ is not a solution of the differential equation $\frac{dy}{dt} + 2y^2 = 3t^2 + y$ because if $y(t) = t^2$, the left hand side of the equation becomes the function $2t + 2t^4$ while the right side of the equation becomes $3t^2 + t^2 = 4t^2$ and we have that these two functions are not the same.

Example 15 *We can check that the function $y(t) = 3\cos(t)$ is not a solution of the differential equation $\frac{dy}{dt} = y + \sin(t)$ because taking y to be $3\cos(t)$, the left hand side of the equation becomes $-3\sin(t)$ while the right hand side of the equation becomes $3\cos(t) + \sin(t)$ and these two functions are not the same.*

Example 16 *We can check that the function $y(t) = 3\mathrm{e}^{2t}$ is a solution of the differential equation $\frac{dy}{dt} = y + 3\mathrm{e}^{2t}$ because taking y to be $3\mathrm{e}^{2t}$, the left hand side of the equation becomes $6\mathrm{e}^{2t}$ while the right hand side of the equation becomes $3\mathrm{e}^{2t} + 3\mathrm{e}^{2t} = 6\mathrm{e}^{2t}$.*

Example 17 *We can check that the function $y(t) = t^2$ is a solution of the differential equation $\frac{dy}{dt} = 2\frac{y}{t}$ because taking y to be t^2, the left hand side of the equation becomes $2t$ while the right hand side of the equation becomes $\frac{2t^2}{t} = 2t$.*

Example 18 *We can check that if A is a constant then, no value of A will make the function $y(t) = A\mathrm{e}^t$ a solution of the differential equation $\frac{dy}{dt} = 2y + \mathrm{e}^{4t}$ because taking y to be $A\mathrm{e}^t$, the left hand side of the equation becomes $A\mathrm{e}^t$ while the right hand side of the equation becomes $2A\mathrm{e}^t + \mathrm{e}^{4t}$.*

Example 19 *We can check that there exists a value of A such that the function $y(t) = A\mathrm{e}^{4t}$ is a solution of the differential equation $\frac{dy}{dt} = 2y + \mathrm{e}^{4t}$ because taking y to be $A\mathrm{e}^{4t}$, the left hand side of the equation becomes $4A\mathrm{e}^{4t}$ while the right hand side of the equation becomes $2A\mathrm{e}^{4t} + \mathrm{e}^{4t} = (2A + 1)\mathrm{e}^{4t}$. Therefore taking $A = \frac{1}{2}$ will make the equation hold true.*

Remark 2 *Notice that since a differential equation is searching for a function then the equation that needs to be verified is an equality of functions, that is, we need to check that two expressions are the same for all values of t. This can be viewed as something good or something bad. On one hand knowing that a function satisfies a differential equation is knowing that this function satisfies infinitely many equations. On the other hand, if we want to find a solution of a differential equation, essentially we need to solve a system with infinitely many equations, one for every t in the domain of the function. This is part of the difficulty of solving differential equations.*

Example 20 *The equation $\frac{dy}{dt} = y$ is a differential equation of order 1. We can check that the function $y(t) = 0$ for all t is a solution of this differential equation as well as the function $y(t) = \mathrm{e}^t$.*

> **Definition 5** *Given a differential equation, we say that a collection of solutions $y(t, c_1, \ldots, c_n)$ parametrized with the constants $c_1, \ldots, c_n$ is a general solution of the differential equation, if for every solution $\tilde{y}$ of the differential solution we can find exactly one choice of constants $c_1, \ldots, c_n$ such that $\tilde{y} = y(t, c_1, \ldots, c_n)$.*

Notice that $y = t^2 + c$ is a general solution of the differential equation $\frac{dy}{dt} = 2t$. Notice that if we consider the solution $\tilde{y} = t^2 + 5$ we have that this solution corresponds with the choice $c = 5$

in the family $y = t^2 + c$. We also have that $y = t^2 - 3c$ is also a general solution, for example, if we can see that $\tilde{y} = t^2 + 5$ is part of this family of solutions with the choice $c = -\frac{5}{3}$. On the other hand $y = t^2 + c_1 - c_2$ is **not** a general solution of $\frac{dy}{dt} = 2t$ because for example for the solution $\tilde{y} = t^2 + 5$ the choice $c_1 = 5$ and $c_2 = 0$ represents $\tilde{y} = t^2 + 5$ and also the choice $c_1 = 6$ and $c_2 = 1$ represents $\tilde{y} = t^2 + 5$. Remember that we are requesting that the choice of the c_i's must be unique. Finally $y = t^2 + c^2$ is not a general solution due to two reasons. On one hand for the solution $\tilde{y} = t^2 + 5$ we have two choices of c, $c = \sqrt{5}$ and $c = -\sqrt{5}$, that represents this particular solution. Also, for the solution $\tilde{y} = t^2 - 1$, no choice of c in the family $y = t^2 + c^2$ represents this solution.

> **Proposition 4** *Given a real number k, the general solution of the differential equation $\frac{dy}{dt} = ky$ is $y(t, c) = ce^{kt}$*

Proof Clearly $y(t, c) = ce^{kt}$ is a solution because $y'(t, c) = kce^{kt} = ky(t, c)$. The next step is to show that if $f(t)$ is a solution with $f(t_0) \neq 0$, then $f(t) \neq 0$ for all t. Near t_0 we have that $\frac{d\ln(|f(t)|)}{dt} = \frac{f'(t)}{f(t)} = k$ and therefore $\ln(|f(t)|) = kt + d$ for some constant d and therefore $|f(t)| = e^d e^{kt}$. By the continuity of $f(t)$ we have that either $f(t) = e^d e^{kt}$ or $f(t) = -e^d e^{kt}$ and therefore $f(t)$ never vanishes. The previous argument shows that either $f(t)$ is identically zero or $f(t) = ce^{kt}$ for some $c \neq 0$. By allowing c to be any real number, including zero, we obtain that any solution of the differential equation corresponds to one function in the collection of functions $y(t, c)$. $\blacksquare$

1.3 Homework

1. Find the real and imaginary part of $(3 - 2i)e^{4t+2it}$. *Answer:* Real part: $2e^{4t} \sin(2t) + 3e^{4t} \cos(2t)$, imaginary part: $3e^{4t} \sin(2t) - 2e^{4t} \cos(2t)$

2. Find the real and imaginary part of $(1 + 3i)e^{-t+2it}$. *Answer:* Real part: $e^{-t} \cos(2t) - 3e^{-t} \sin(2t)$, imaginary part: $e^{-t} \sin(2t) + 3e^{-t} \cos(2t)$

3. Find the real and imaginary part of $(-1 + i)e^{-4it}$. *Answer:* Real part: $\sin(4t) - \cos(4t)$, imaginary part: $\sin(4t) + \cos(4t)$

4. Find the real and imaginary part of $(2i)e^{2t+3it}$.

   Answer: Real part: $-2e^{2t} \sin(3t)$, imaginary part: $2e^{2t} \cos(3t)$

5. Find the solution of the equation $x^2 + 3x + 3 = 0$.

   Answer: $x = \frac{1}{2}\left(-3 - i\sqrt{3}\right)$, $x = \frac{1}{2}\left(-3 + i\sqrt{3}\right)$

6. Find the solution of the equation $x^2 + 2x - 3 = 0$.

   Answer: $x = -3$, $x = 1$

7. Find the solution of the equation $x^2 + 4x + 4 = 0$.

   Answer: $x = -2$

8. Find the solution of the equation $x^2 - 5x + 2 = 0$.

   Answer: $x = \frac{1}{2}\left(5 - \sqrt{17}\right)$, $x = \frac{1}{2}\left(\sqrt{17} + 5\right)$

9. Find the solution of the equation $x^2 - 2x + 5 = 0$.

   Answer: $x = 1 - 2i$, $x = 1 + 2i$

10. Find the solution of the equation $x^2 - 6x + 9 = 0$. *Answer:* $x = 3$

11. Check that $z = 2 + 4i$ is a solution of the equation $x^2 - 4x + 20 = 0$. *Hint:* $z^2 = -12 + 16i$

12. Check that $z = -1 + 3i$ is a solution of the equation $x^2 + 2x + 10 = 0$. *Hint:* $z^2 = -8 - 6i$

13. Check that $z = 2 + \sqrt{5}i$ is a solution of the equation $x^2 - 4x + 9 = 0$. *Hint:* $z^2 = -1 + 4i\sqrt{5}$

14. Compute the first and second derivative of the function $f(t) = e^{2t}\sin(3t)$. *Answer:* $\frac{df}{dt} = 2e^{2t}\sin(3t) + 3e^{2t}\cos(3t)$, $\frac{d^2f}{dt^2} = 12e^{2t}\cos(3t) - 5e^{2t}\sin(3t)$.

15. Compute the first and second derivative of the function $f(t) = 3e^{-2t}\sin(5t) + e^{-2t}\cos(5t)$.
    Answer: $\frac{df}{dt} = 13e^{-2t}\cos(5t) - 11e^{-2t}\sin(5t)$, $\frac{d^2f}{dt^2} = -43e^{-2t}\sin(5t) - 81e^{-2t}\cos(5t)$.

16. Compute the first and second derivative of the function $f(t) = 3e^{3t}\sin\left(\sqrt{2}t\right) + e^{3t}\cos\left(\sqrt{2}t\right)$.
    Answer: $\frac{df}{dt} = -\sqrt{2}e^{3t}\sin\left(\sqrt{2}t\right) + 9e^{3t}\sin\left(\sqrt{2}t\right) + 3e^{3t}\cos\left(\sqrt{2}t\right) + 3\sqrt{2}e^{3t}\cos\left(\sqrt{2}t\right)$,
    $\frac{d^2f}{dt^2} = -6\sqrt{2}e^{3t}\sin\left(\sqrt{2}t\right) + 21e^{3t}\sin\left(\sqrt{2}t\right) + 7e^{3t}\cos\left(\sqrt{2}t\right) + 18\sqrt{2}e^{3t}\cos\left(\sqrt{2}t\right)$.

17. Decide if $y(t) = e^{2t}$ is a solution of the differential equation $y'(t) + e^{2t} = 2y(t)$. *Answer* No, because the left hand side is equal to $3e^{2t}$ while the right hand side is $2e^{2t}$.

18. Decide if $y(t) = -2te^{2t}$ is a solution of the differential equation $y'(t) + 2e^{2t} = 2y(t)$. *Answer* Yes, because the left hand side and the right hand side are equal to $-4te^{2t}$.

19. Decide if $y(t) = -e^{5t}$ is a solution of the differential equation $y'(t) + 4e^{5t} = y(t)$. *Answer* Yes, because the left and the right hand side are equal to $-e^{5t}$.

20. Decide if $y(t) = 3$ is a solution of the differential equation $y'(t) + 3y(t) = 6$. *Answer* No, because the left hand side is equal to $y'(t) + 3y(t) = 9$ while the right hand side is the constant function 6.

21. Decide if $y(t) = t$ is a solution of the differential equation $y'(t) + y(t) = t$. *Answer* No, because the left hand side is equal to $y'(t) + y(t) = 1 + t$ while the right hand side is t.

22. Find a and b such that $y(t) = at + b$ is a solution of $y'(t) + 2y(t) = t$. *Answer* $a = 1/2$ and $b = -1/4$.

23. Simplify the expression $e^{2\ln(t)}$; assume $t > 0$. Answer t^2.

24. Simplify the expression $e^{\ln(2t)}$; assume $t > 0$. Answer $2t$.

25. Simplify the expression $e^{-\ln(t^2+1)}$. Answer $\frac{1}{t^2+1}$.

26. Simplify the expression $e^{\frac{1}{2}\ln(3t^4+1)}$. Answer $\sqrt{3t^4 + 1}$.

27. Find the general solution of the differential equation $\frac{dy}{dt} = 2y$. Answer $y = ce^{2t}$

28. Find the general solution of the differential equation $\frac{dy}{dt} + 3y = 0$. Answer $y = ce^{-3t}$

29. Find the general solution of the differential equation $2\frac{dy}{dt} + 5y = 0$. Answer $y = ce^{-\frac{5}{2}t}$

chapter 2

Analytic techniques to solve some differential equations

It is fair to say that most differential equations cannot be explicitly solved. In this section we study some classes of differential equations for which there is a method to find a solution.

We have already explained the notion of a general solution. We now explain the notion of an initial value problem.

Definition 6 *Given a differential equation of order n, an initial value problem searches for a function that satisfies the differential equation as well as n conditions of the form* $y(t_0) = y_0, \ldots, \frac{d^{n-1}y}{dt^{n-1}}(t_0) = y_{n-1}.$

Example 21 *The solution of the initial value problem $\frac{dy}{dt} = 2t$, $y(3) = 4$ is* $\boxed{y(t) = t^2 - 5}$.

Example 22 *The solution of the initial value problem $\frac{d^2y}{dt^2} + 4y = 0$, $y(0) = 1$, $y'(0) = 2$ is the function* $\boxed{y(t) = \cos(2t) + \sin(2t)}$.

Given the general solution $y(t, c_1, \ldots, c_n)$ we can use the conditions of the initial value problem to find the right constants $c_1, \ldots, c_n$ that give us the solution of the initial value problem.

2.1 Differential equations of the form $\frac{dy}{dt} = f(t)$

In this section we emphasize that finding antiderivatives are particular examples of solving differential equations. We have,

Proposition 5 *The general solution of the differential equation $\frac{dy}{dt} = f(t)$ is $y(t) = F(t) + c$ where $F(t)$ is an antiderivative of $f(t)$.*

Example 23 *Let us solve the initial value problem $\frac{dy}{dt} = \sqrt{t}$, $y(1) = 3$. We have that*

$$\int \sqrt{t}\,dt = \int t^{1/2}\,dt = \frac{2}{3}t^{3/2}$$

Therefore the general solution of the differential equation is $y(t) = \frac{2}{3}t^{3/2} + c$. Now, the initial condition $y(1) = 3$ reduces to the equation $\frac{2}{3}1^{3/2} + c = \frac{2}{3} + c = 3$. Therefore $c = 3 - \frac{2}{3} = \frac{7}{3}$ and the solution of the initial value problem is $\boxed{y(t) = \dfrac{2}{3}t^{3/2} + \dfrac{7}{3}}$.

Example 24 *Let us solve the initial value problem $\frac{dy}{dt} = -\frac{6}{t^2}$, $y(2) = 1$. We have that*

$$\int \frac{-6}{t^2}\,dt = \int -6t^{-2}\,dt = 6t^{-1} = \frac{6}{t}$$

Therefore the general solution of the differential equation is $y(t) = \frac{6}{t} + c$. Now, the initial condition $y(2) = 1$ reduces to the equation $\frac{6}{2} + c = 3 + c = 1$. Therefore $c = -2$ and the solution of the initial value problem is $\boxed{y(t) = \dfrac{6}{t} - 2}$.

Example 25 *Let us solve the initial value problem $\frac{dy}{dt} = \frac{2}{1+t^2}$, $y(0) = 0$. We have that*

$$\int \frac{2}{1 + t^2}\,dt = 2\arctan(t)$$

Therefore the general solution of the differential equation is $y(t) = 2\arctan(t) + c$. Now, the initial condition $y(0) = 0$ reduces to the equation $2\arctan(0) + c = 0$. Therefore $c = 0$ and the solution of the initial value problem is $\boxed{y(t) = 2\arctan(t)}$.

2.2 Equilibrium Solutions

Some differential equations have some solutions of the form $y(t) = A$ where A is a constant. These types of solutions are called **equilibrium solutions**, and when they exist, they play an important role in the understanding of other solutions of the differential equation.

> To find out if a differential equation has equilibrium solutions we just check if $y(t) = A$ solves the equation for any value or values of A. Notice that since A is constant then $\frac{dy}{dt}$ as well as all the higher derivatives of y are the zero function.

Therefore the problem reduces to that of solving an algebraic equation for the variable A, with A a real number. When $y(t) = A$ is an equilibrium solution we say that A is an **equilibrium point**.

Example 26 *To find the equilibrium solutions of the differential equation $\frac{dy}{dt} = y^2 - 3y + 2$ we replace $y(t) = A$ to obtain the equation $0 = A^2 - 3A + 2 = (A - 1)(A - 2) = 0$. Therefore, this differential equation has two equilibrium solutions $\boxed{y(t) = 1}$ and $\boxed{y(t) = 2}$.*

Example 27 *The differential equation $\frac{dy}{dt} = y + t$ does not have equilibrium solutions because when we replace $y(t) = A$ we obtain the equation $0 = A + t$ which does not have a solution. Notice that we are assuming that A is constant and therefore the answer $A = -t$ is not a valid answer. Recall that we already have used that A is constant when we replaced $\frac{dy}{dt}$ with zero.*

Example 28 *Replacing $y(t) = A$ into the differential equation $\frac{dy}{dt} = t(y - 2)$ we obtain that $0 = t(A - 2)$. This equation holds true for every t if $A = 2$. Therefore $\boxed{y(t) = 2}$ is an equilibrium solution.*

2.3   Separation of variables

This method is an application of the chain rule.

> **Proposition 6** *If $G(u)$ is an antiderivative of $g(u)$ and $F(t)$ is an antiderivative of $f(t)$ then the general solution of the differential equation $g(y)\frac{dy}{dt} = f(t)$ is the collection of functions $y(t)$ found by solving for $y(t)$ from the equations $G(y(t)) = F(t) + c$.*

Proof The proof follows from the fact that the differential equation tells us the derivative of the function

$$G(y(t)) - F(t)$$

is zero and therefore this function must be a constant c.

> **Remark 3** *This method is called separation of variables because if we pretend that $\frac{dy}{dt}$ is the quotient of dy with dt then the differential equation $g(y)\frac{dy}{dt} = f(t)$ can be written as $g(y)dy = f(t)dt$. That is, the differential equation can be written as an expression with no t or dt on one side, and an expression with no y or dy on the other side of the equation.*

Example 29 *Let us solve the initial value problem $\frac{dy}{dt} = 2ty$ with $y(0) = 2$. We have that the differential equation can be written as*

$$\frac{1}{y}\,dy = 2t\,dt \tag{2.3.1}$$

Therefore we need to find the antiderivative of $\frac{1}{y}$ and the antiderivative of $2t$. (in this case $g(u) = \frac{1}{u}$ and $f(t) = 2t$. We can skip this reasoning and just say that we integrate both parts of Equation (2.3.1). We obtain that $\ln|y| = t^2 + c$ and therefore $|y| = e^c e^{t^2}$ or equivalently $y = \pm e^c e^{t^2}$. Since $y(t) = 0$ is an equilibrium solution and $\pm e^c$ can be any nonzero real number, we get that the general solution of the differential equation is $y = c_1 e^{t^2}$. Using the initial condition we obtain that $y(0) = 2 = c_1 e^0 = c_1$. Therefore the solution of the initial value problem is $\boxed{y(t) = 2e^{t^2}}$

Example 30 *Let us solve the initial value problem $\frac{dy}{dt} = 2 - 3y$ with $y(0) = -1$. We have that the differential equation can be written as*

$$\frac{1}{2 - 3y}\, dy = dt \tag{2.3.2}$$

We obtain that $-\frac{1}{3}\ln|2 - 3y| = t + c$ and therefore $|2 - 3y| = e^c e^{-3t}$ or equivalently $2 - 3y = \pm e^c e^{-3t}$. Since $y(t) = 2/3$ is an equilibrium solution and $\pm e^c$ can be any nonzero real number we get that the general solution of the differential equation is $y = \frac{2}{3} + c_1 e^{-3t}$. Using the initial condition we obtain that $y(0) = -1 = \frac{2}{3} + c_1$, this is $c_1 = -\frac{5}{3}$. Therefore the solution of the initial value problem is $\boxed{y(t) = \dfrac{2}{3} - \dfrac{5}{3}e^{-3t}}$.

Example 31 *Let us solve the initial value problem $\frac{dy}{dt} = \frac{2t+1}{y}$ with $y(1) = -2$. We have that the differential equation can be written as*

$$y\,dy = (2t + 1)\, dt \tag{2.3.3}$$

We obtain that $\frac{y^2}{2} = t^2 + t + c$. This time it is more convenient to find the constant before solving for y. Replacing $y(1) = -2$ we obtain the equation $\frac{(-2)^2}{2} = 1 + 1 + c$ and therefore $c = 0$. Solving for y we obtain that $y(t) = \pm\sqrt{2t^2 + 2t}$. We need to decide if we need the plus or the minus, we do this by looking again at the initial condition. The solution of the initial value problem is $\boxed{y(t) = -\sqrt{2t^2 + 2t}}$.

Example 32 *Let us solve the initial value problem $\frac{dy}{dt} = \frac{e^t}{2y+1}$ with $y(0) = -4$. We have that the differential equation can be written as*

$$(2y + 1)\, dy = e^t\, dt \tag{2.3.4}$$

We obtain that $y^2 + y = e^t + c$. Replacing $y(0) = -4$ we obtain the equation $(-4)^2 + (-4) = 1 + c$ and therefore $c = 11$. In order to solve for y we set up the equation equal to zero and we use the quadratic formula. We obtain that

$$y = \frac{-1 \pm \sqrt{1 - 4(-e^t - 11)}}{2} = \frac{-1 \pm \sqrt{45 + 4e^t}}{2}$$

Since we want $y(0) = -4$ then the solution of the initial value problem is $\boxed{y(t) = \dfrac{-1 - \sqrt{45 + 4e^t}}{2}}$.

Example 33 *Let us solve the initial value problem $\frac{dy}{dt} = -2y^2 t + 4y^2$ with $y(1) = 2$. We have that the differential equation can be written as*

$$\frac{1}{y^2}\, dy = (-2t + 4)\, dt \tag{2.3.5}$$

We obtain that $-\frac{1}{y} = -t^2 + 4t + c$. *Replacing* $y(1) = 2$ *we obtain the equation* $-\frac{1}{2} = -1 + 4 + c$ *and therefore* $c = -\frac{7}{2}$. *We now solve for* y

$$-\frac{1}{y} = -t^2 + 4t - \frac{7}{2} \quad \text{or equivalently} \quad -\frac{1}{-t^2 + 4t - \frac{7}{2}} = y$$

therefore, the solution of the initial value problem is $\boxed{y(t) = \dfrac{1}{t^2 - 4t + \frac{7}{2}}}$.

Example 34 *Let us solve the initial value problem* $\frac{dy}{dt} = \frac{yt}{1+t^2}$ *with* $y(0) = 1$. *We have that the differential equation can be written as*

$$\frac{1}{y}\, dy = \frac{t}{1 + t^2}\, dt \tag{2.3.6}$$

We obtain that $\ln|y| = \frac{1}{2}\ln(1 + t^2) + c$. *In order to solve the integral* $\int \frac{t}{1+t^2}\, dt$ *we have used the substitution* $u = 1 + t^2$, *then* $du = 2t\, dt$ *and therefore* $\int \frac{t}{1+t^2}\, dt = \frac{1}{2}\int \frac{1}{u}\, du$. *From the equation* $\ln|y| = \frac{1}{2}\ln(1 + t^2) + c$ *we obtain that*

$$|y| = e^c e^{\frac{1}{2}\ln(1+t^2)} = e^c(e^{\ln(1+t^2)})^{\frac{1}{2}} = e^c(1 + t^2)^{\frac{1}{2}} = e^c\sqrt{1 + t^2}$$

and then $y = \pm e^c\sqrt{1 + t^2}$. *Since* $y = 0$ *is a solution and* $\pm e^c$ *can be any non zero number, then we obtain that the general solution is* $y = c_1\sqrt{1 + t^2}$. *Replacing* $y(0) = 1$ *we obtain that* $c_1 = 1$.

Therefore, the solution of the initial value problem is $\boxed{y(t) = \sqrt{1 + t^2}}$.

2.4 Homework

Solve the following initial value problems.

1. Find the equilibrium solutions if any: $\frac{dy}{dt} = y^2 - t$, Answer: None

2. Find the equilibrium solutions if any: $\frac{dy}{dt} = t(y^2 - 4)$, Answer: $y = 2$ and $y = -2$

3. Find the equilibrium solutions if any: $\frac{dy}{dt} = (y^2 - 4y + 4)$, Answer: $y = 2$

4. Find the equilibrium solutions if any: $\frac{dy}{dt} = y(y^2 - 2y - 3)$, Answer: $y = 0$, $y = -1$ and $y = 3$

5. Find the equilibrium solutions if any: $\frac{dy}{dt} = (y - 2)(y^2 - 5y + 20)$, Answer: $y = 2$

6. Find the equilibrium solutions if any: $\frac{dy}{dt} = (y - 2)(y^2 - 5y - 3)$, Answer: $y = 2$, $y = \frac{1}{2}\left(5 - \sqrt{37}\right)$ and $y = \frac{1}{2}\left(\sqrt{37} + 5\right)$

7. Find the general solution $\frac{dy}{dt} = 3$, Answer: $y = 3t + c$

8. Find the general solution $\frac{dy}{dt} = \frac{5}{2+3t}$, Answer: $y = \frac{5}{3}\ln|2+3t| + c$

9. Find the general solution $\frac{dy}{dt} = \frac{2}{1+4t^2}$, Answer: $y = \arctan(2t) + c$

10. Find the general solution $\frac{dy}{dt} = \frac{2}{(t+3)(t+5)}$, Answer: $y = \ln(3+t) - \ln(t+5) + c$

11. $\frac{dy}{dt} = y^2(\sin(t) + e^{2t})$, $y(0) = 1$. Answer $y = -\frac{2}{e^{2t} - 2\cos(t) - 1}$.

12. $\frac{dy}{dt} = y(4t+1)$, $y(1) = -2$. Answer $y = -2e^{2t^2+t-3}$.

13. $\frac{dy}{dt} = yt$, $y(0) = -5$. Answer $y = -5e^{\frac{t^2}{2}}$.

14. $\frac{dy}{dt} = y^3\cos(t)$, $y(0) = -3$. Answer $y = -\frac{3}{\sqrt{1-18\sin(t)}}$.

15. $\frac{dy}{dt} = 3y + 2$, $y(0) = -2$. Answer $y = -\frac{4e^{3t}}{3} - \frac{2}{3}$.

16. $\frac{dy}{dt} = 2 - 5y$, $y(0) = 1$. Answer $y = \frac{3e^{-5t}}{5} + \frac{2}{5}$.

17. $\frac{dy}{dt} = 4y^2t$, $y(1) = 2$. Answer $y = -\frac{2}{4t^2-5}$.

18. $\frac{dy}{dt} = \frac{4t+2}{2y+1}$, $y(1) = -3$. Answer $y = -\frac{1}{2}\sqrt{8t^2 + 8t + 9} - \frac{1}{2}$.

19. $\frac{dy}{dt} = \frac{\sin(t)}{2y+3}$, $y(0) = 0$. Answer $y = \frac{1}{2}\sqrt{13 - 4\cos(t)} - \frac{3}{2}$.

20. $\frac{dy}{dt} = \frac{3t^2}{y(t^3+1)}$, $y(0) = 6$. Answer $y = \sqrt{2}\sqrt{\ln(t^3 + 1) + 18}$.

21. $\frac{dy}{dt} = \frac{2yt}{t^2+1}$, $y(0) = -3$. Answer $y = -3t^2 - 3$.

22. $\frac{dy}{dt} = \frac{yt}{t^2+1}$, $y(0) = -3$. Answer $y = -3\sqrt{t^2 + 1}$.

23. $\frac{dy}{dt} = (y^2 + 1)$, $y(0) = 0$. Answer $y = \tan(t)$.

24. $\frac{dy}{dt} = (y^2 + 1)2t$, $y(0) = 0$. Answer $y = \tan(t^2)$.

chapter 3

Order one linear equations

Let us start with the definition of linear differential equation of order one.

> **Definition 7** *A linear equation of order one is a differential equation that can be written as*
>
> $$\frac{dy}{dt} + g(t)y = b(t) \qquad (3.0.1)$$
>
> *If $b(t)$ is zero we say that the linear equation is a homogeneous equation. When $b(t)$ is not zero we say that the linear equation is non homogeneous. We say that*
>
> $$\frac{dy}{dt} + g(t)y = 0$$
>
> *is the homogeneous equation associated with*
>
> $$\frac{dy}{dt} + g(t)y = b(t)$$

Example 35 *The differential equation $(2t^2+7)\frac{dy}{dt}+2y = \sin(t)$ is linear. Notice that after dividing the equation by $2t^2+7$, the equation can be written $\frac{dy}{dt} + \frac{2}{2t^2+7}y = \frac{\sin(t)}{2t^2+7}$ which has the form given in the definition with $g(t) = \frac{2}{2t^2+7}$ and $b(t) = \frac{\sin(t)}{2t^2+7}$*

Example 36 *The differential equation $y\frac{dy}{dt} = \sin(t)$ is not linear, neither is the equation $\frac{dy}{dt} + \sin(y) = 4$.*

The following theorem will be useful finding the general solution of a linear differential equation.

> **Theorem 2** *If $y_p(t)$ is a solution of $\frac{dy}{dt}+g(t)y = b(t)$, and $y_H(t,c)$ is the general solution of $\frac{dy}{dt} + g(t)y = 0$, then the general solution of $\frac{dy}{dt}+g(t)y = b(t)$ is $y(t,c) = y_p(t)+y_H(t,c)$.*

Proof Let us first check that for any c, $y = y_p(t)+y_H(t,c)$ solves the non homogeneous differential equation. We have that

$$\frac{dy}{dt} + g(t)y = \frac{dy_p}{dt} + \frac{dy_H}{dt} + g(t)(y_p + y_H) = \frac{dy_p}{dt} + g(t)y_p + \frac{dy_H}{dt} + g(t)y_H = b(t)$$

On the other hand if $y(t)$ is a solution of $\frac{dy}{dt} + g(t)y = b(t)$, it is easy to check that $y(t) - y_p(t)$ is a solution of $\frac{dy}{dt} + g(t)y = 0$. Therefore, for some c, $y(t) - y_p(t) = y_H(t, c)$ and $y(t) = y_p(t) + y_H(t, c)$.

Example 37 *We can easily check that $y_p(t) = 3$ is a solution of the equation $\frac{dy}{dt} = 2y - 6$ because $y = 3$ is an equilibrium solution. We also have the general solution of $\frac{dy}{dt} = 2y$ is $y_H(t, c) = ce^{2t}$. Therefore the general solution of $\frac{dy}{dt} = 2y - 6$ is* $\boxed{y(t, c) = 3 + ce^{2t}}$.

3.1 Guessing technique.

When $g(t)$ is constant the solution of the homogeneous is easy to find, it is of the form $y_H(t, c) = ce^{kt}$. For the particular solution we look for $y_p(t)$ according to $b(t)$. We have:

1. If $b(t) = c\sin(kt)$ or $b(t) = c\cos(kt)$ or $b(t) = c_1 \sin(kt) + c_2 \cos(kt)$, then we look for $y_p(t) = A\cos(kt) + B\sin(kt)$. Notice that even if $b(t)$ is just $3\sin(2t)$ we have to make $y_p(t) = A\cos(2t) + B\sin(2t)$.

2. If $b(t)$ is a polynomial of order n, then we look for $y_p(t) = A_0 + \cdots + A_n t^n$. Notice that, even if $b(t) = 4t^3$ we still need to try $y_p(t) = A + Bt + Ct^2 + Dt^3$.

3. If $b(t) = de^{kt}$ we look for $y_p(t) = Ae^{kt}$ unless the solution of the homogeneous is $y_H(t, c) = ce^{kt}$, in this case, we look for $y_p(t) = Ate^{kt}$. This latter choice for $y_p(t)$ is called the second guess.

Example 38 *Let us find the general solution of $\frac{dy}{dt} = 3y + 4e^{2t}$. Here the associated homogeneous equation is $\frac{dy}{dt} = 3y$ and therefore $y_H(t, c) = ce^{3t}$. We look for the particular solution of the form $y_p(t) = Ae^{2t}$. The unknown A must satisfy*

$$2Ae^{2t} = 3(Ae^{2t}) + 4e^{2t} \quad \text{or equivalently} \quad 0 = (A + 4)e^{2t}$$

Therefore $A = -4$ and $y_p(t) = -4e^{2t}$ is a particular solution and the general solution of $\frac{dy}{dt} = 3y + 4e^{2t}$ is $\boxed{y(t, c) = -4e^{2t} + ce^{3t}}$.

Example 39 *Let us find the general solution of $\frac{dy}{dt} = 3y + 4e^{3t}$. Here the associated homogeneous equation is $\frac{dy}{dt} = 3y$ and therefore $y_H(t, c) = ce^{3t}$. Our first guess is to look for the particular solution of the form $y_p(t) = Ae^{3t}$, but since $y_H(t, c) = ce^{3t}$, then we need to use the second guess and try to find A such that $y_p(t) = Ate^{3t}$. The unknown A must satisfy*

$$Ae^{3t} + 3Ate^{3t} = 3(Ate^{3t}) + 4e^{3t} \quad \text{or equivalently} \quad (A - 4)e^{3t} = 0 \tag{3.1.1}$$

Therefore $A = 4$ and $y_p(t) = 4te^{3t}$ is a particular solution and the general solution of $\frac{dy}{dt} = 3y + 4e^{3t}$ is $\boxed{y(t, c) = 4te^{3t} + ce^{3t}}$. *Notice that in Equation (3.1.1) the terms that contains Ate^{3t} cancel out. This always happens when we are using a second guess.*

Example 40 *Let us find the general solution of $\frac{dy}{dt} + y = t + 4e^t$. Here the associated homogeneous equation is $\frac{dy}{dt} = -y$ and therefore $y_H(t, c) = ce^{-t}$. We look for the particular solution of the form $y_p(t) = Ae^t + Bt + C$. The unknowns A, B and C must satisfy*

$$Ae^t + B + (Ae^t + Bt + C) = t + 4e^t \quad \text{or equivalently} \quad (2A - 4)e^t + (B - 1)t + B + C = 0$$

The equation above holds true if $A - 2 = 0$, $B - 1 = 0$ and $B + C = 0$. Therefore $A = 2$, $B = 1$ and $C = -1$, and $y_p(t) = 2e^t + t - 1$ is a particular solution and the general solution of $\frac{dy}{dt} + y = t + 4e^t$ is $\boxed{y(t, c) = 2e^t + t - 1 + ce^{-t}}$.

Example 41 *Let us solve the initial value problem $\frac{dy}{dt} - 2y = 3\sin 2t$, $y(0) = 2$. Here the associated homogeneous equation is $\frac{dy}{dt} = 2y$ and therefore $y_H(t, c) = ce^{2t}$. We look for the particular solution of the form $y_p(t) = A\sin 2t + B\cos 2t$. The unknowns A and B must satisfy*

$$2A\cos 2t - 2B\sin 2t - 2(A\sin 2t + B\cos 2t) = 3\sin 2t \quad \text{or equivalently}$$

$$(2A - 2B)\cos 2t + (-2A - 2B - 3)\sin 2t = 0$$

The equation above holds true if $2A - 2B = 0$ and $-2A - 2B - 3 = 0$. Therefore $A = -\frac{3}{4}$ and $B = -\frac{3}{4}$ and $y_p(t) = -\frac{3}{4}\sin 2t - \frac{3}{4}\cos 2t$ is a particular solution and the general solution of is $y(t, c) = -\frac{3}{4}\sin 2t - \frac{3}{4}\cos 2t + ce^{2t}$. Since we want $y(0) = 2$, then $-\frac{3}{4} + c = 2$ and $c = \frac{11}{4}$. Then the solution of the initial value problem is $\boxed{y(t) = \frac{11}{4}e^{2t} - \frac{3}{4}\sin 2t - \frac{3}{4}\cos 2t}$

Example 42 *Let us solve the general solution of $\frac{dy}{dt} - 2y = 3t^2$. Here the associated homogeneous equation is $\frac{dy}{dt} = 2y$ and therefore $y_H(t, c) = ce^{2t}$. We look for the particular solution of the form $y_p(t) = A + Bt + Ct^2$. The unknowns A, B and C must satisfy*

$$B + 2Ct - 2(A + Bt + Ct^2) = 3t^2$$

or equivalently

$$(B - 2A) + (2C - 2B)t + (-2C - 3)t^2 = 0$$

The equation above holds true if $B - 2A = 0$, $2C - 2B = 0$ and $-2C - 3 = 0$. Therefore $C = -\frac{3}{2}$, $B = -\frac{3}{2}$ and $A = -\frac{3}{4}$. Therefore $y_p(t) = -\frac{3}{4} - \frac{3}{2}t - \frac{3}{2}t^2$ is a particular solution and the general solution of is $\boxed{y(t, c) = ce^{2t} - \frac{3}{4} - \frac{3}{2}t - \frac{3}{2}t^2}$.

Example 43 *Let us solve the solution of the initial value problem $\frac{dy}{dt} + 2y = 4e^{-2t} + t - 3\cos(2t)$ with $y(0) = 10$. Here the associated homogeneous equation is $\frac{dy}{dt} + 2y = 0$ and therefore $y_H(t, c) = ce^{-2t}$. Our first guess for a particular solution has the form $y_p(t) = A + Bt + De^{-2t} + E\cos(2t) + F\sin(2t)$. We notice that the part De^{-2t} is a multiple of the solution of the homogeneous. Therefore this part needs to be replaced by the second guess. We therefore try the function*

$$y_p(t) = A + Bt + Dte^{-2t} + E\cos(2t) + F\sin(2t)$$

Replacing this function in the differential equation we get that

$$(B+De^{-2t}-2Dte^{-2t}-2E\sin(2t)+2F\cos(2t))+2(A+Bt+Dte^{-2t}+E\cos(2t)+F\sin(2t))$$

must be equal to

$$4e^{-2t}+t-3\cos(2t)$$

Simplifying we get the following expression must vanish

$$B+2A+2Bt+De^{-2t}+(2F-2E)\sin(2t)+(2E+2F)\cos(2t)=4e^{-2t}+t-3\cos(2t),$$

Therefore, the unknowns A, B, D, E and F must satisfy

$$B+2A=0,\ 2B=1,\ D=4,\ 2F-2E=0,\quad and \quad 2E+2F=-3$$

The solution of the system is $B=1/2$, $D=4$, $A=-1/4$, $F=-3/4$ and $E=-3/4$. Therefore the general solution is

$$y=y_H+y_p=ce^{-2t}-1/4+1/2t+4te^{-2t}-3/4\cos(2t)-3/4\sin(2t)$$

Now we use the initial condition $y(0)=10$ to find c. We have that $y(0)=c-1/4-3/4=10$. Therefore $c=11$ and the solution of the initial value problem is

$$\boxed{y(t)=11e^{-2t}-1/4+1/2t+4te^{-2t}-3/4\cos(2t)-3/4\sin(2t)}.$$

3.2 Solving the general case using integrating factor

Let us start by pointing out that $\frac{de^{t^2}}{dt}=2te^{t^2}$, $\frac{de^{\sin t}}{dt}=(\cos t)e^{\sin t}$ and in general $\frac{de^{u(t)}}{dt}=u'(t)e^{u(t)}$. The previous observation is the key step of the following theorem.

> **Theorem 3** *If $G'(t)=g(t)$ and $B'(t)=e^{G(t)}f(t)$, then the general solution of the differential equation $\frac{dy}{dt}+g(t)y=b(t)$ is $y(t,c)=e^{-G(t)}(B(t)+c)$.*

Proof By multiplying the equation $\frac{dy}{dt}+g(t)y=b(t)$ by $e^{G(t)}$ we obtain that

$$e^{G(t)}\frac{dy}{dt}+g(t)e^{G(t)}y=e^{G(t)}b(t) \tag{3.2.1}$$

The left hand side of Equation (3.2.1) is the derivative of $e^{G(t)}y(t)$ while the right hand side is the derivative of $B(t)$. Therefore the function $e^{G(t)}y(t)-B(t)$ is constant and the result follows.

> **Remark 4** *According to the previous theorem these are the steps that we need to do to solve a linear differential equation.*
>
> 1. *Rewrite the linear equation if needed so that we identify $g(t)$ and $b(t)$.*
>
> 2. *Find an antiderivative of $g(t)$ and call it $G(t)$*
>
> 3. *Compute $\mu(t) = e^{G(t)}$. Make sure to simplify if possible. Recall the property $e^{a \ln(b)} = b^a$. The function μ is called the integrating factor.*
>
> 4. *Compute an antiderivative of $\mu b(t)$ and call this function $B(t)$.*
>
> 5. *The general solution is $y(t, c) = \frac{1}{\mu(t)}(B(t) + c)$*

Example 44 *Let us find the general solution of $\frac{dy}{dt} = \frac{y}{t} + 3$. In this case $g(t) = -\frac{1}{t}$ and $b(t) = 3$. Therefore $G(t) = -\ln(t)$, $\mu = e^{-\ln(t)}$ which simplifies to $\frac{1}{t}$ and $B(t) = \int \frac{3}{t}\,dt = 3\ln(t)$ and therefore the general solution $\boxed{y(t,c) = t\,(3\ln(t) + c)}$*

Example 45 *Let us find the general solution of $\frac{dy}{dt} + \frac{y}{1+t} = 6t$. In this case $g(t) = \frac{1}{1+t}$ and $b(t) = 6t$. Therefore $G(t) = \ln(1+t)$, $\mu = e^{\ln(1+t)}$ which simplifies to $1+t$ and $B(t) = \int (1+t)6t\,dt = 3t^2 + 2t^3$ and therefore the general solution $\boxed{y(t,c) = \frac{1}{1+t}\left(3t^2 + 2t^3 + c\right) = \frac{3t^2 + 2t^3 + c}{1+t}}$*

Example 46 *Let us find the general solution of $\frac{dy}{dt} - 2ty = 3t^2 e^{t^2}$. In this case $g(t) = -2t$ and $b(t) = 3t^2 e^{t^2}$. Therefore $G(t) = -t^2$, $\mu = e^{-t^2}$. Now $B(t) = \int e^{-t^2} 3t^2 e^{t^2}\,dt = \int 3t^2\,dt = t^3$ and therefore the general solution $\boxed{y(t,c) = e^{t^2}\left(t^3 + c\right)}$*

Example 47 *Let us find the general solution of $\frac{dy}{dt} + \frac{2ty}{1+t^2} = 3$. In this case $g(t) = \frac{2t}{1+t^2}$ and $b(t) = 3$. Therefore $G(t) = \int \frac{2t}{1+t^2}\,dt = \ln(1+t^2)$. In the previous integral we have used substitution using $u = 1+t^2$ and $du = 2t\,dt$. We obtain that $\mu = 1+t^2$. Now $B(t) = \int 3(1+t^2)\,dt = 3t + t^3$ and therefore the general solution $\boxed{y(t,c) = \frac{t^3 + 3t + c}{1+t^2}}$*

Example 48 *Let us find the general solution of $\frac{dy}{dt} - \frac{1}{1+t}y = 3$. In this case $g(t) = -\frac{1}{1+t}$ and $b(t) = 3$. Therefore $G(t) = -\int \frac{1}{1+t}\,dt = -\ln(1+t)$. In the previous integral we have used substitution using $u = 1+t$ and $du = dt$. We obtain that $\mu = \frac{1}{1+t}$. Now $B(t) = \int \frac{3}{1+t}\,dt = 3\ln(1+t)$ and therefore the general solution $y = \frac{1}{\mu}(3\ln(1+t) + c)$, this is $\boxed{y(t,c) = (1+t)\,(3\ln(1+t) + c)}$*

Example 49 *Let us find the solution of the initial value problem $\frac{dy}{dt} = -2ty + 4e^{-t^2}$ with $y(0) = 3$. In this case $g(t) = 2t$ and $b(t) = 4e^{-t^2}$. Therefore $G(t) = \int 2t\,dt = t^2$. We obtain that $\mu = e^{t^2}$. Now $B(t) = \int e^{t^2} 4e^{-t^2}\,dt = \int 4\,dt = 4t$ and therefore the general solution $y = e^{-t^2}(4t + c)$. Using the initial condition $y(0) = 3$ we obtain that $y(0) = 1(0 + c) = c = 3$, and the solution of the initial value problem is $\boxed{y(t,c) = e^{-t^2}(4t + 3)}$.*

Example 50 *Let us find the general solution of $\frac{dy}{dt} - \frac{2t}{1+t^2}y = 3$. In this case $g(t) = -\frac{2t}{1+t^2}$ and $b(t) = 3$. Therefore $G(t) = -\int \frac{2t}{1+t^2}\,dt = -\ln(1+t^2)$. In the previous integral we have used substitution using $u = 1 + t^2$ and $du = 2t\,dt$. We obtain that $\mu = \frac{1}{1+t^2}$. Now $B(t) = \int \frac{3}{1+t^2}\,dt = 3\arctan(t)$ and therefore the general solution $y = (1+t^2)(3\arctan(t) + c$.*

$$\boxed{y(t,c) = (1+t^2)\left(3\arctan(t) + c\right)}.$$

3.3 Homework

For each one of the following problems. Solve the initial value problem and then evaluate the solution at $t = 0.3$ using 5 significant digits.

1. $\frac{dy}{dt} = 3y + 5$, $y(0) = 2$. Answer: $y = \frac{11e^{3t}}{3} - \frac{5}{3}$, 7.35188.

2. $\frac{dy}{dt} = -2y + 5t$, $y(0) = -4$. Answer: $y = \frac{5t}{2} - \frac{11e^{-2t}}{4} - \frac{5}{4}$, -2.00923.

3. $\frac{dy}{dt} + 4y = 3e^{2t}$, $y(0) = -1$. Answer: $y = \frac{1}{2}(-3)e^{-4t} + \frac{e^{2t}}{2}$, 0.459268.

4. $\frac{dy}{dt} - 6y = 2e^{2t}$, $y(0) = 2$. Answer: $y = \frac{5e^{6t}}{2} - \frac{e^{2t}}{2}$, 14.2131.

5. $\frac{dy}{dt} = 3y + 5e^{3t}$, $y(0) = 2$. Answer: $y = 5e^{3t}t + 2e^{3t}$, 8.60861.

6. $\frac{dy}{dt} + y = 2e^{-t}$, $y(0) = 6$. Answer: $y = 2e^{-t}t + 6e^{-t}$, 4.8894.

7. $\frac{dy}{dt} - 2y = 2\sin(2t)$, $y(0) = 1$. Answer: $y = \frac{3e^{2t}}{2} - \frac{1}{2}\sin(2t) - \frac{1}{2}\cos(2t)$, 2.03819.

8. $\frac{dy}{dt} + 2y = 2\sin(2t) - \cos(2t)$, $y(0) = 0$. Answer: $y = \frac{3e^{-2t}}{4} + \frac{1}{4}\sin(2t) - \frac{1}{4}3\cos(2t)$, -0.0662324.

9. $\frac{dy}{dt} = -y + 4t^2$, $y(0) = 3$. Answer: $y = 4t^2 - 8t - 5e^{-t} + 8$, 2.25591.

10. $\frac{dy}{dt} = 2y + e^{2t} + 4t^2$, $y(0) = 0$. Answer: $y = -2t^2 + e^{2t}t - 2t + e^{2t} - 1$, 0.588754.

11. $\frac{dy}{dt} = -\frac{2}{1+t}y + 3t$, $y(0) = 0$. Answer: $y = \frac{t^2\left(3t^2+8t+6\right)}{4(t+1)^2}$, 0.115429.

12. $\frac{dy}{dt} + \frac{2}{t}y = 3$, $y(1) = 2$. Answer: $y = \frac{1}{t^2} + t$, 11.4111.

13. $\frac{dy}{dt} + \frac{4}{t}y = 2t$, $y(1) = 2$. Answer: $y = \frac{t^6+5}{3t^4}$, 205.791.

14. $\frac{dy}{dt} - \frac{2t}{1+t^2}y = 2$, $y(0) = -1$. Answer: $y = (t^2 + 1)\left(2\tan^{-1}(t) - 1\right)$, -0.454624.

15. $\frac{dy}{dt} + \frac{2t}{1+t^2}y = 2$, $y(0) = -1$. Answer: $y = \frac{2t^3+6t-3}{3(t^2+1)}$, -0.350459.

chapter 4

Linear second order differential equations

In this section we will study the differential equation of the form

$$\frac{d^2y}{dt^2} + b\frac{dy}{dt} + cy = F(t) \tag{4.0.1}$$

when $F(t)$ is a linear combination of a polynomial, an exponential, and sine and/or cosine functions.

This case is very similar to the first order linear equation case where we use the first guess and the second guess. We have,

Theorem 4 *If $y_p(t)$ is a solution of $\frac{d^2y}{dt^2} + b\frac{dy}{dt} + cy = F(t)$, and $y_H(t, c_1, c_2)$ is the general solution of $\frac{d^2y}{dt^2} + b\frac{dy}{dt} + cy = 0$, then the general solution of $\frac{d^2y}{dt^2} + b\frac{dy}{dt} + cy = F(t)$ is $y(t, c_1, c_2) = y_p(t) + y_H(t, c_1, c_2)$.*

4.1 Solution of the homogeneous

The following theorem explains how to find the general solution $y_H(t, c_1, c_2)$.

Theorem 5 *The general solution of the homogeneous linear equation $\frac{d^2y}{dt^2} + b\frac{dy}{dt} + cy = 0$ is*

1. *$y_H(t, c_1, c_2) = c_1 e^{\lambda_1 t} + c_2 t e^{\lambda_1 t}$ if $b^2 = 4c$, where λ_1 is the only solution of the equation $\lambda^2 + b\lambda + c = 0$.*

2. *$y_H(t, c_1, c_2) = c_1 e^{\lambda_1 t} + c_2 e^{\lambda_2 t}$ if $b^2 - 4c > 0$, where λ_1 and λ_2 are the two real solutions of $\lambda^2 + b\lambda + c = 0$.*

3. *$y_H(t, c_1, c_2) = c_1 e^{\alpha t} \cos(\beta t) + c_2 e^{\alpha t} \sin(\beta t)$ if $b^2 - 4c < 0$, where $\beta > 0$ and $\alpha \pm \beta i$ are the two solutions of the equation $\lambda^2 + b\lambda + c = 0$*

4.2 Finding the particular solution

For the particular solution we look for $y_p(t)$ according to the non-homogeneous part $F(t)$. We have:

1. If $F(t) = c\sin(kt)$ or $F(t) = c\cos(kt)$ or $F(t) = c_1\sin(kt) + c_2\cos(kt)$, then we look for $y_p(t) = A\cos(kt) + B\sin(kt)$, unless $y_H = c_1\cos(kt) + c_2\sin(kt)$; in this case we look for the second guess $y_p(t) = At\cos(kt) + Bt\sin(kt)$. We call this latter case, **resonance**.

2. In general we proceed as in the order one case. We take a first guess similar to the non homogeneous part $F(t)$ and if this guess is a multiple of a part of y_H, then we multiply this guess by t. We call this the second guess. It may happen that the second guess is also a multiple of some part of the y_H and therefore we need to multiply the second guess by t.

4.3 Examples

Example 51 *Compute the solution of the initial value problem $\frac{d^2y}{dt^2} + 3\frac{dy}{dt} - 4y = 2e^t$ with $y(0) = 1$ and $y'(0) = 4$. In this case, the homogeneous equation is $\frac{d^2y}{dt^2} + 3\frac{dy}{dt} - 4y = 0$. Since $b^2 - 4ac = 3^2 - 4(-4) = 25 > 0$, then the equation $\lambda^2 + 3\lambda - 4 = 0$ has two real solutions, they are $\lambda_1 = 1$ and $\lambda_2 = -4$. Using Theorem 5 we get that $y_H(t, c_1, c_2) = c_1 e^t + c_2 e^{-4t}$. The first guess for the particular solution is $y_p = Ae^t$ but since this is a multiple of part of the solution y_H we need to consider the second guess $y_p = Ate^t$. We notice that this second guess is not a multiple of any part of the solution of y_H, thus it will work. We have that $\frac{dy_p}{dt} = Ae^t + Ate^t$ and $\frac{d^2y_p}{dt^2} = 2Ae^t + Ate^t$. Since we want y_p to solve the non-homogeneous equation we must have*

$$(2Ae^t + Ate^t) + 3(Ae^t + Ate^t) - 4Ate^t = 2e^t$$

As expected the terms with te^t cancel out and the equation reduces to $5Ae^t = 2e^t$ which is true if $A = \frac{2}{5}$. Therefore $y_p = \frac{2}{5}te^t$ and the general solution is

$$y = \frac{2}{5}te^t + c_1 e^t + c_2 e^{-4t}$$

In order to find the initial conditions we need to take the derivative of the general solution.

$$y' = \frac{2}{5}e^t + \frac{2}{5}te^t + c_1 e^t - 4c_2 e^{-4t}$$

Replacing $y(0) = 1$ and $y'(0) = 4$ in the expression for y and y' above, give us the following system

$$c_1 + c_2 = 1 \quad and \quad \frac{2}{5} + c_1 - 4c_2 = 4$$

the solution of the system is $c_1 = \frac{38}{25}$ and $c_2 = -\frac{13}{25}$. Therefore the solution of the initial value problem is $\boxed{y = \dfrac{2e^t t}{5} - \dfrac{13}{25}e^{-4t} + \dfrac{38}{25}e^t}$

Example 52 *Compute the solution of the initial value problem $\frac{d^2y}{dt^2} + 2\frac{dy}{dt} + y = 3\cos t$ with $y(0) = 2$ and $y'(0) = 0$. In this case, the homogeneous equation is $\frac{d^2y}{dt^2} + 2\frac{dy}{dt} + y = 0$. Since $b^2 - 4ac = 2^2 - 4(1) = 0$, then the equation $\lambda^2 + 2\lambda + 1 = 0$ has only one solution, $\lambda_1 = -1$. Using Theorem 5 we get that $y_H(t, c_1, c_2) = c_1 e^{-t} + c_2 t e^{-t}$. The first guess for the particular solution is $y_p = A\cos t + B\sin t$. Since this guess is not a multiple of any part of the solution of y_H, it will work. We have that $\frac{dy_p}{dt} = -A\sin t + B\cos t$ and $\frac{d^2y_p}{dt^2} = -A\cos t - B\sin t$. Since we want y_p to solve the non-homogeneous equation we must have*

$$(-A\cos t - B\sin t) + 2(-A\sin t + B\cos t) + A\cos t + B\sin t = 3\cos t$$

The equation above can be written as $2B\cos t - 2A\sin t = 3\cos t$. Comparing coefficients of the functions $\sin(t)$ and $\cos t$ we obtain that $B = \frac{3}{2}$ and $A = 0$ and $y_p = \frac{3}{2}\sin t$. Therefore the general solution is

$$y = \frac{3}{2}\sin t + c_1 e^{-t} + c_2 t e^{-t}$$

In order to find the initial conditions we need to take the derivative of the general solution.

$$y' = \frac{3}{2}\cos t - c_1 e^{-t} + c_2 e^{-t} - c_2 t e^{-t}$$

Replacing $y(0) = 2$ and $y'(0) = 0$ in the expression for y and y' above, give us the following system

$$c_1 = 2 \quad and \quad \frac{3}{2} - c_1 + c_2 = 0$$

the solution of the system is $c_1 = 2$ and $c_2 = \frac{1}{2}$. Therefore the solution of the initial value problem is $\boxed{y = \dfrac{3}{2}\sin t + 2e^{-t} + \dfrac{1}{2}t e^{-t}}$

Example 53 *Compute the solution of the initial value problem $\frac{d^2y}{dt^2} + 2\frac{dy}{dt} + 10y = 2e^{-t}$ with $y(0) = 0$ and $y'(0) = 0$. In this case, the homogeneous equation is $\frac{d^2y}{dt^2} + 2\frac{dy}{dt} + 10y = 0$. Since $b^2 - 4ac = 2^2 - 4(10) = -36 < 0$, then the equation $\lambda^2 + 2\lambda + 10 = 0$ has two non-real solutions, they are $\lambda_1 = -1 + 3i$ and $\lambda_2 = -1 - 3i$. Therefore, $\alpha = -1$ and $\beta = 3$. Using Theorem 5 we get that $y_H(t, c_1, c_2) = c_1 e^{-t}\cos(3t) + c_2 e^{-t}\sin(3t)$. The first guess for the particular solution is $y_p = Ae^{-t}$, since this guess is not a multiple of any part of the solution of y_H, then it will work. We would like to emphasize that e^{-t} is not a multiple of $e^{-t}\sin(3t)$ even if they share an e^{-t}. We have that $\frac{dy_p}{dt} = -Ae^{-t}$ and $\frac{d^2y_p}{dt^2} = Ae^{-t}$. Since we want y_p to solve the non-homogeneous equation we must have*

$$Ae^{-t} + 2(-Ae^{-t}) + 10Ae^{-t} = 2e^{-t}$$

the equation reduces to $9Ae^{-t} = 2e^{-t}$ which is true if $A = \frac{2}{9}$. Therefore $y_p = \frac{2}{9}e^{-t}$ and the general solution is

$$y = \frac{2}{9}e^{-t} + c_1 e^{-t}\cos(3t) + c_2 e^{-t}\sin(3t)$$

In order to find the initial conditions we need to take the derivative of the general solution.

$$y' = -\frac{2}{9}e^{-t} - c_1 e^{-t}\cos(3t) - 3c_1 e^{-t}\sin(3t) - c_2 e^{-t}\sin(3t) + 3c_2 e^{-t}\cos(3t)$$

Replacing $y(0) = 0$ and $y'(0) = 0$ in the expression for y and y' above, give us the following system

$$\frac{2}{9} + c_1 = 0 \quad and \quad -\frac{2}{9} - c_1 + 3c_2 = 0$$

the solution of the system is $c_1 = -\frac{2}{9}$ and $c_2 = 0$. Therefore the solution of the initial value problem is $\boxed{y = \frac{2}{9}e^{-t} - \frac{2}{9}e^{-t}\cos(3t)}$

Example 54 *Compute the solution of the initial value problem $\frac{d^2y}{dt^2} + 4y = 2t$ with $y(0) = 1$ and $y'(0) = -1$. In this case, the homogeneous equation is $\frac{d^2y}{dt^2} + 4y = 0$. Since $b^2 - 4ac = 0 - 4 \times 4 = -16 < 0$, then the equation $\lambda^2 + 4 = 0$ has two non-real solutions, they are $\lambda_1 = 2i$ and $\lambda_2 = -2i$. In this case $\alpha = 0$ and $\beta = 2$. Using Theorem 5 we get that $y_H(t, c_1, c_2) = c_1\cos(2t) + c_2\sin(2t)$. The first guess for the particular solution is $y_p = At + B$, since this guess is not a multiple of any part of the solution of y_H, then it will work. We have that $\frac{dy_p}{dt} = A$ and $\frac{d^2y_p}{dt^2} = 0$. Since we want y_p to solve the non-homogeneous equation we must have*

$$0 + 4(At + B) = 2t$$

the equation reduces to $4At + 4B = 2t$ which is true if $4A = 2$ and $4B = 0$. Therefore $y_p = \frac{1}{2}t$ and the general solution is

$$y = \frac{1}{2}t + c_1\cos(2t) + c_2\sin(2t)$$

In order to find the initial conditions we need to take the derivative of the general solution.

$$y' = \frac{1}{2} - 2c_1\sin(2t) + 2c_2\cos(2t)$$

Replacing $y(0) = 1$ and $y'(0) = -1$ in the expression for y and y' above, give us the following system

$$c_1 = 1 \quad and \quad \frac{1}{2} + 2c_2 = -1$$

the solution of the system is $c_1 = 0$ and $c_2 = -\frac{3}{4}$. Therefore the solution of the initial value problem is $\boxed{y = \frac{1}{2}t + \cos(2t) - \frac{3}{4}\sin(2t)}$

Example 55 *Compute the general solution of $\frac{d^2y}{dt^2} + 6\frac{dy}{dt} + 9y = 4e^{-3t}$. In this case, the homogeneous equation is $\frac{d^2y}{dt^2} + 6\frac{dy}{dt} + 9y = 0$. Since $b^2 - 4ac = 36 - 4 \times 9 = 0$, then the equation $\lambda^2 + 6\lambda + 9 = 0$ has only one solution, $\lambda_1 = -3$. Using Theorem 5 we get that $y_H(t, c_1, c_2) = c_1 e^{-3t} + c_2 t e^{-3t}$. The first guess for the particular solution is $y_p = Ae^{-3t}$, since this guess is a multiple of the first term of y_H, then we need to try the second guess $y_p = Ate^{-3t}$. Since this second* *guess is a multiple is the second term of y_H, then we need the third guess $y_p = At^2 e^{-3t}$. Since this is not multiple of any part of y_H, then it will work. We have that $\frac{dy_p}{dt} = 2Ate^{-3t} - 3At^2 e^{-3t}$ and $\frac{d^2 y_p}{dt^2} = 2Ae^{-3t} - 6Ate^{-3t} - 6Ate^{-3t} + 9At^2 e^{-3t}$. Since we want y_p to solve the non-homogeneous equation we must have*

$$(2Ae^{-3t} - 12Ate^{-3t} + 9At^2 e^{-3t}) + 6(2Ate^{-3t} - 3At^2 e^{-3t}) + 9At^2 e^{-3t} = 4e^{-3t}$$

the equation reduces to $2Ae^{-3t} = 4Ae^{-3t}$ which is true if $2A = 4$. Therefore $y_p = 2t^2 e^{-3t}$ and the general solution is $\boxed{y = 2t^2 e^{-3t} + c_1 e^{-3t} + c_2 t e^{-3t}}$

Example 56 *Solve the initial value problem $\frac{d^2 y}{dt^2} + 4y = 2\sin(2t)$, $y(0) = 0$, $y'(0) = 0$. In order to find y_H we notice that the solution of the equation $\lambda^2 + 4 = 0$ is $\lambda = \pm 2i$ and therefore $y_H(t, c_1, c_2) = c_1 \cos(2t) + c_2 \sin(2t)$. For the particular solution we try $y_p = A\cos(2t) + B\sin(2t)$, but after noticing that this is a multiple of y_H we realize that we need a second guess and we try $y_p = At\cos(2t) + Bt\sin(2t)$. We have*

$$\frac{dy_p}{dt} = A\cos(2t) - 2At\sin(2t) + B\sin(2t) + 2Bt\cos(2t)$$

and

$$\frac{d^2 y_p}{dt^2} = -4A\sin(2t) - 4At\cos(2t) + 4B\cos(2t) - 4Bt\sin(2t)$$

Therefore, we want the following equation to be true

$$\frac{d^2 y_p}{dt^2} + 4y_p = -4A\sin(2t) + 4B\cos(2t) = 2\sin(2t)$$

We see that $A = -\frac{1}{2}$ and $B = 0$ solves the equation. The general solution becomes

$$y(t, c_1, c_2) = c_1 \cos(2t) + c_2 \sin(2t) - \frac{1}{2} t \cos(2t)$$

and in order to find the constant c_1 and c_2 we compute

$$\frac{dy}{dt} = -2c_1 \sin(2t) + 2c_2 \cos(2t) - \frac{1}{2}\cos(2t) + t\sin(2t)$$

and replace the initial conditions to obtain the following system of equations

$$c_1 = 0, \quad 2c_2 - \frac{1}{2} = 0$$

Then, the solution of the initial value problem is $\boxed{y = \frac{1}{4}\sin(2t) - \frac{1}{2} t \cos(2t)}$

The previous example is an example of resonance. More precisely we have the following definition.

Definition 8 *We say that the differential equation $\frac{d^2 y}{dt^2} + \omega^2 y = A\cos(\omega t) + B\sin(\omega t)$ is at resonance due to the fact that a second guess is needed for the solution and therefore the solution oscillates and is unbounded.*

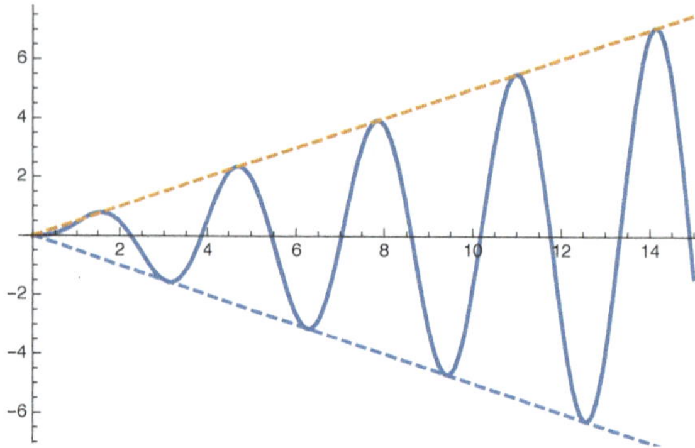

Figure 4.3.1: In the case of resonance the solution is unbounded. It grows between two lines oscillating from one line to another. It is not fun to move along this solution if you easily get dizzy.

4.4   Near resonance

The differential equation $\frac{d^2y}{dt^2} + \omega^2 y = 0$ has as a solution the harmonic motion $y = c_1\cos(\omega t) + c_2\sin(\omega t)$. For this homogeneous equation we will call ω the **the natural angular frequency**. As pointed out in example (56), if we add a non homogeneous part of the form $\sin(\omega t)$ or $\cos(\omega t)$ then we will need a second guess (making resonance show up) and our solution will be unbounded. In this section we will learn to do a sketch of the solution when the non homogeneous part is of the form $\sin(\omega_1 t)$ or $\cos(\omega_1 t)$ with ω_1 close to ω. We call this case, **near resonance**.

Let us consider the following initial value problem

$$\frac{d^2y}{dt^2} + \omega^2 y = \cos(\omega_1 t), \quad y(0) = y'(0) = 0 \quad \text{with } \omega_1 \text{ near } \omega \tag{4.4.1}$$

We look for the particular of the form $y_p = A\cos(\omega_1 t) + B\sin(\omega_1 t)$ and we obtain that $(\omega^2 - \omega_1^2)A = 1$ and $B = 0$. Then the general solution is

$$y = c_1\cos(\omega t) + c_2\sin(\omega t) + \frac{1}{\omega^2 - \omega_1^2}\cos(\omega_1 t)$$

Using the initial conditions we get that the solution of the initial value problem is

$$y = \frac{1}{\omega^2 - \omega_1^2}\left(\cos(\omega_1 t) - \cos(\omega t)\right) \tag{4.4.2}$$

Some basic trigonometry identities will allow us to rewrite the solution above. We have that

$$\cos(x \pm y) = \cos(x)\cos(y) \mp \sin(x)\sin(y)$$
$$\sin(x \pm y) = \sin(x)\cos(y) \pm \sin(y)\cos(x)$$

Using the first equation above we conclude that $\cos(x+y) - \cos(x-y) = -2\sin(x)\sin(y)$ and making $x + y = \omega_1 t$ and $x - y = \omega t$ we get that $x = \frac{1}{2}(\omega + \omega_1)t$ and $y = \frac{1}{2}(\omega_1 - \omega)t$. Therefore, the expression in Equation (4.4.2) can be written as

$$y = \frac{2}{\omega^2 - \omega_1^2} \sin(\frac{1}{2}(\omega + \omega_1)t) \sin(\frac{1}{2}(\omega - \omega_1)t) \tag{4.4.3}$$

An advantage of the last form of writing the solution is that we have that the function y crosses the t axis anytime any of the two functions $\sin(\frac{1}{2}(\omega + \omega_1)t)$ and $\sin(\frac{1}{2}(\omega - \omega_1)t)$ cross the t axis.

Remember that the function $\sin(\Omega t)$ has period $P = \frac{2\pi}{\Omega}$ and in the closed interval $[0, P]$ the function $\sin(\Omega t)$ vanishes at the end points and in the middle point.

The previous observations allow us to have a sketch of the solution of the differential equation (4.4.1) by doing the following steps.

1. Identify the natural angular frequency ω and the forcing angular frequency ω_1.

2. Compute $\omega_b = \frac{1}{2}|\omega - \omega_1|$ and $\omega_{ro} = \frac{1}{2}|\omega + \omega_1|$.

3. Compute the periods associated with the two angular frequencies computed in the previous step. This is $P_{ro} = \frac{2\pi}{\omega_{ro}}$ and $P_b = \frac{2\pi}{\omega_b}$.

4. Draw with dashed lines the graphs of $y = \sin(\omega_b t)$ and $y = -\sin(\omega_b t)$. This part of the motion is called the **beats**.

5. Divide the real line into intervals with length P_{ro} and in each one of these intervals draw a sine-like function making sure that the graph crosses the t axis at the beginning, at the end and in the middle of each interval. The sine-like function must go up as much as possible while staying inside the beats. This going up and down in a sinusoidal way over these small intervals is called **rapid oscillations**.

Example 57 *Let us sketch the solution of the differential equation $\frac{d^2y}{dt^2} + 9y = 5\cos(\frac{5t}{2})$. Step one is to point out the natural angular frequency is $\omega = 3$ and the forcing angular frequency is $\omega_1 = \frac{5}{2}$. For step 2 we compute $\omega_b = \frac{1}{2}(3 - \frac{5}{2}) = \frac{1}{4}$ and $\omega_{ro} = \frac{1}{2}(3 + \frac{5}{2}) = \frac{11}{4}$. For step 3 we compute the periods using the formula $P = \frac{2\pi}{\omega}$. The period of the beats is $P_b = 8\pi$ and the period of the rapid oscillations is $P_{ro} = \frac{8\pi}{11}$. The sketch of the graph is shown in Figure 4.4.1*

Example 58 *Let us sketch the solution of the differential equation $\frac{d^2y}{dt^2} + 25y = \cos(4t)$. The natural angular frequency is $\omega = 5$ and the forcing angular frequency is $\omega_1 = 4$. Now we compute the other two angular frequencies $\omega_b = \frac{1}{2}(5 - 4) = \frac{1}{2}$ and $\omega_{ro} = \frac{1}{2}(4 + 5) = \frac{9}{2}$. Therefore, the period of the beats is $P_b = 4\pi$ and the period of the rapid oscillations is $P_{ro} = \frac{4\pi}{9}$. The graph is shown in Figure 4.4.2*

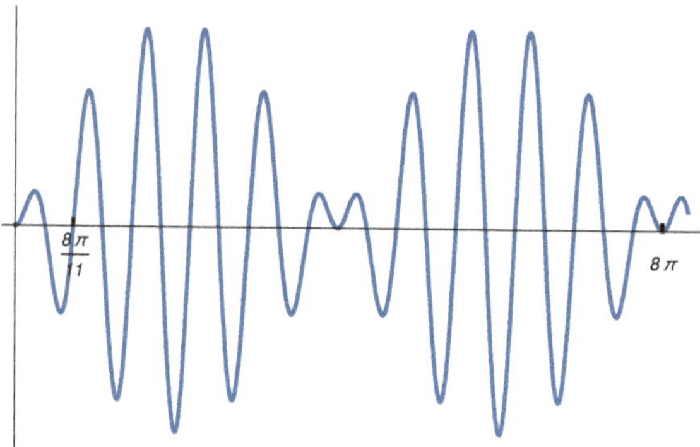

Figure 4.4.1: Notice that we have labeled the period of the rapid oscillations and the period of the beats

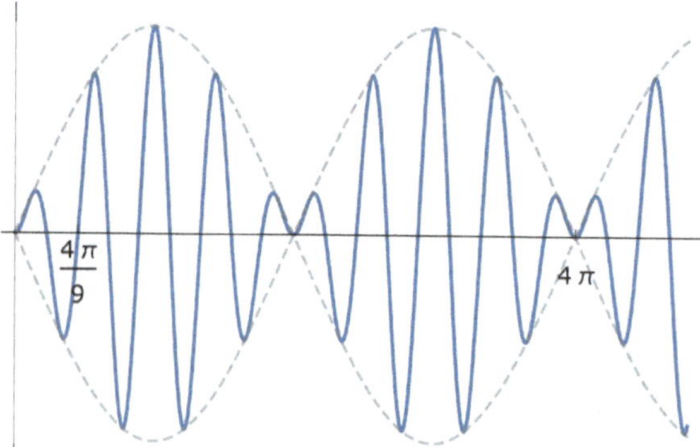

Figure 4.4.2: The graphs of the beats, the dashed curves, are not part of the graph, they just help us to do the sketch of the solution of the ODE.

4.5 Homework

1. Solve the initial value problem $\frac{d^2y}{dt^2} + \frac{dy}{dt} - 6y = 3e^{4t}$, $y(0) = -1$, $y'(0) = 2$, Answer: $y = \frac{1}{7}(-5)e^{-3t} - \frac{e^{2t}}{2} + \frac{3e^{4t}}{14}$.

2. Solve the initial value problem $\frac{d^2y}{dt^2} + 2\frac{dy}{dt} + 2y = 2\sin(2t)$, $y(0) = 0$, $y'(0) = 1$, Answer: $y = \frac{9}{5}e^{-t}\sin(t) - \frac{1}{5}\sin(2t) + \frac{2}{5}e^{-t}\cos(t) - \frac{2}{5}\cos(2t)$.

3. Solve the initial value problem $\frac{d^2y}{dt^2} - 4\frac{dy}{dt} + 4y = 3e^{5t}$, $y(0) = -2$, $y'(0) = 1$, Answer: $y = 4e^{2t}t + \frac{e^{5t}}{3} - \frac{7e^{2t}}{3}$.

4. Solve the initial value problem $\frac{d^2y}{dt^2} - 3\frac{dy}{dt} + 2y = 3e^{2t}$, $y(0) = 2$, $y'(0) = 1$, Answer: $y = 3e^{2t}t + 6e^t - 4e^{2t}$.

5. Solve the initial value problem $\frac{d^2y}{dt^2} - \frac{dy}{dt} - 6y = 3e^{2t}$, $y(0) = -2$, $y'(0) = 1$, Answer: $y = -\frac{5}{4}e^{-2t} - \frac{3e^{2t}}{4}$.

6. Solve the initial value problem $\frac{d^2y}{dt^2} + 4\frac{dy}{dt} + 5y = 4e^t$, $y(0) = -2$, $y'(0) = 4$, Answer: $y = \frac{2e^t}{5} - \frac{6}{5}e^{-2t}\sin(t) - \frac{12}{5}e^{-2t}\cos(t)$.

7. Solve the initial value problem $\frac{d^2y}{dt^2} - 6\frac{dy}{dt} + 13y = 5\sin 2t$, $y(0) = 0$, $y'(0) = 4$, Answer: $y = \frac{11}{5}e^{3t}\sin(2t) + \frac{1}{5}\sin(2t) - \frac{1}{15}4e^{3t}\cos(2t) + \frac{4}{15}\cos(2t)$.

8. Solve the initial value problem $\frac{d^2y}{dt^2} - 6\frac{dy}{dt} + 9y = 2e^{3t}$, $y(0) = -2$, $y'(0) = -1$, Answer: $y = e^{3t}t^2 + 5e^{3t}t - 2e^{3t}$.

9. Solve the initial value problem $\frac{d^2y}{dt^2} - 4\frac{dy}{dt} + 3y = 2 + 3t$, $y(0) = -1$, $y'(0) = 2$, Answer: $y = t - 5e^t + 2e^{3t} + 2$.

10. Solve the initial value problem $\frac{d^2y}{dt^2} - 2\frac{dy}{dt} + y = 5\cos(3t)$, $y(0) = -1$, $y'(0) = 0$, Answer: $y = \frac{3e^t t}{2} - \frac{3e^t}{5} - \frac{3}{10}\sin(3t) - \frac{2}{5}\cos(3t)$.

11. Solve the initial value problem $\frac{d^2y}{dt^2} + 4y = 5\cos(3t)$, $y(0) = -1$, $y'(0) = 0$, Answer: $y = -\cos(3t)$

12. Solve the initial value problem $\frac{d^2y}{dt^2} + 100y = 19\cos(9t)$, $y(0) = 0$, $y'(0) = 0$, Answer: $y = \cos(9t) - \cos(10t)$

13. Solve the initial value problem $\frac{d^2y}{dt^2} - 4\frac{dy}{dt} + 8y = 32t^2$, $y(0) = 2$, $y'(0) = -4$, Answer: $y = 4t^2 + 4t - 5e^{2t}\sin(2t) + e^{2t}\cos(2t) + 1$

14. Solve the initial value problem $\frac{d^2y}{dt^2} - 2\frac{dy}{dt} = e^t$, $y(0) = 2$, $y'(0) = -4$, Answer: $y = -e^t - \frac{3e^{2t}}{2} + \frac{9}{2}$

15. Solve the initial value problem $\frac{d^2y}{dt^2} + 4y = 8\sin(2t)$, $y(0) = 2$, $y'(0) = -4$, Answer: $y = -\sin(2t) - 2t\cos(2t) + 2\cos(2t)$

16. Solve the initial value problem $\frac{d^2y}{dt^2} + 9y = 3\cos(3t)$, $y(0) = 0$, $y'(0) = 0$, Answer: $y = \frac{1}{2}t\sin(3t)$

17. Solve the initial value problem $\frac{d^2y}{dt^2} - 4\frac{dy}{dt} + 3y = 2e^t$, $y(0) = -1$, $y'(0) = 2$, Answer: $y = -e^t t - 3e^t + 2e^{3t}$

18. Solve the initial value problem $\frac{d^2y}{dt^2} = 12t^2$, $y(0) = -1$, $y'(0) = 2$, Answer: $y = t^4 + 2t - 1$

19. Consider $\frac{d^2y}{dt^2} + 4y = 4\cos(\frac{5}{2}t)$ $y(0) = 0$, $y'(0) = 0$. Sketch the graph of the solution making sure to label the period of the beats and the period of the rapid oscillations.

20. Consider $\frac{d^2y}{dt^2} + 64y = 4\cos(7t)$ $y(0) = 0$, $y'(0) = 0$. Sketch the graph of the solution making sure to label the period of the beats and the period of the rapid oscillations.

21. Consider $\frac{d^2y}{dt^2} + 6y = 4\cos(3t)$ $y(0) = 0$, $y'(0) = 0$. Sketch the graph of the solution making sure to label the period of the beats and the period of the rapid oscillations.

22. Consider $\frac{d^2y}{dt^2} + 25y = 4\cos(4t)$ $y(0) = 0$, $y'(0) = 0$. Sketch the graph of the solution making sure to label the period of the beats and the period of the rapid oscillations.

23. Consider $\frac{d^2y}{dt^2} + 4y = 4\cos(\frac{8}{5}t)$ $y(0) = 0$, $y'(0) = 0$. Sketch the graph of the solution making sure to label the period of the beats and the period of the rapid oscillations.

24. Consider $\frac{d^2y}{dt^2} + 10y = 4\cos(\frac{5}{2}t)$ $y(0) = 0$, $y'(0) = 0$. Sketch the graph of the solution making sure to label the period of the beats and the period of the rapid oscillations.

25. Consider $\frac{d^2y}{dt^2} + 36y = 4\cos(\frac{11}{2}t)$ $y(0) = 0$, $y'(0) = 0$. Sketch the graph of the solution making sure to label the period of the beats and the period of the rapid oscillations.

Answers problems 19 to 25:

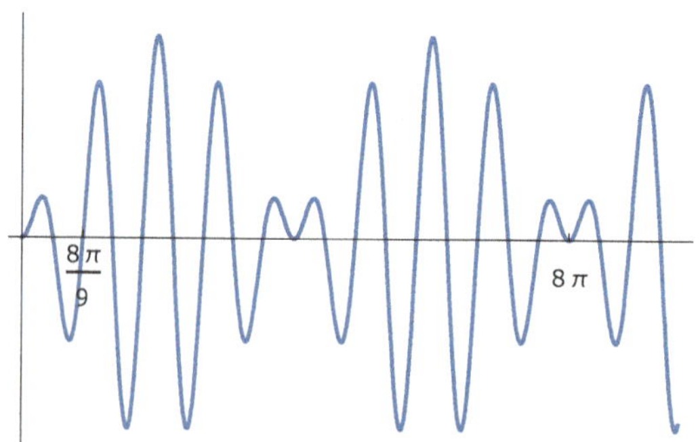

Figure 4.5.1: Answer problem 19

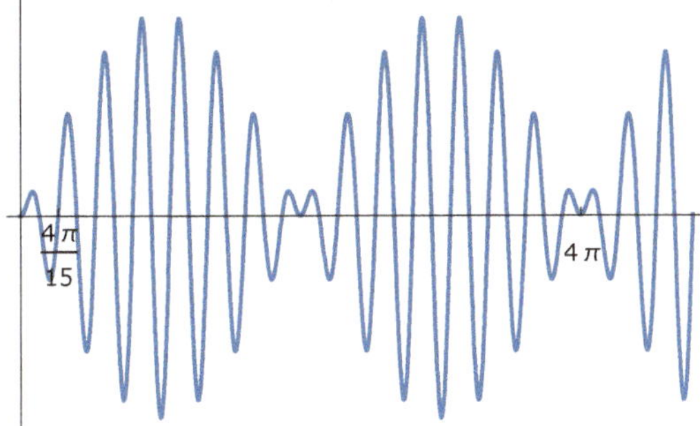

Figure 4.5.2: Answer problem 20

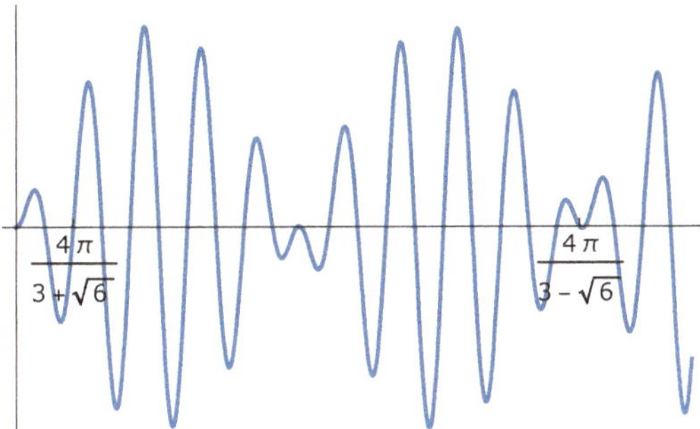

Figure 4.5.3: Answer problem 21

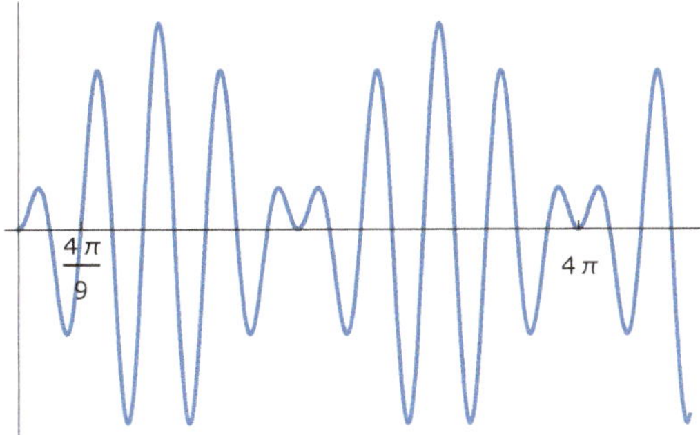

Figure 4.5.4: Answer problem 22

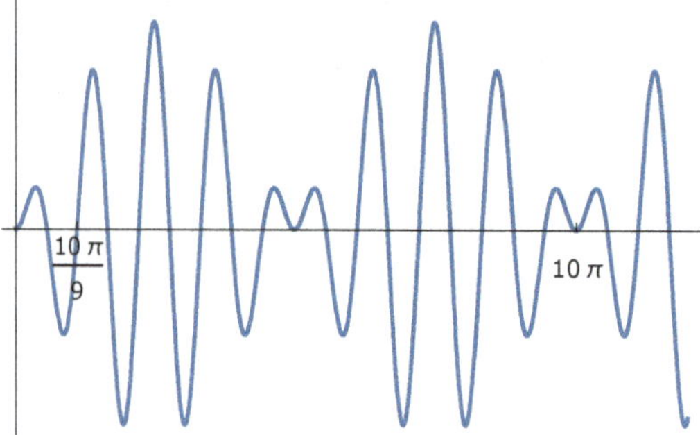

Figure 4.5.5: Answer problem 23

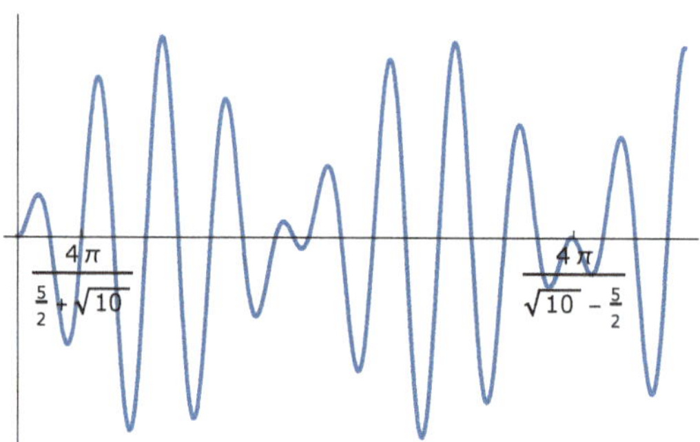

Figure 4.5.6: Answer problem 24

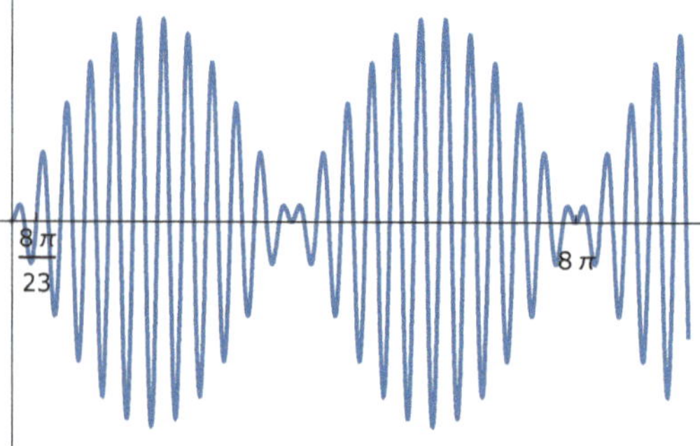

Figure 4.5.7: Answer problem 25

chapter 5

Phase line

5.1 Definition

The phase line of a differential equation allows us to get some information about the solutions of a differential equation without having to explicitly solve it. It shows the importance of computing/studying the equilibrium points of a differential equation.

> **Definition 9** *If $f(y)$ is a smooth function defined for all real numbers that only vanishes at the points $y_1, \ldots y_n$, then the **phase line** of the differential equation $\frac{dy}{dt} = f(y)$ is a vertical line with dots at the points y_i and with arrows between the dots that point up if f is positive and point down if f is negative. The equilibrium points y_i's is called a **sink** if the arrow before and after point towards it. It is called a source if the arrows before and after point away from it and it is called a **node** if the arrows before and after both point up or both point down. If the function is alway positive then the phase line is just a vertical line with an arrow going up. Likewise, if the function $f(y)$ is always negative, then the phase line is just a line with an arrow going down.*

Example 59 *Let us find the phase line of the differential equation $\frac{dy}{dt} = y^2(y^2 - 1)$. For this example $f(y) = y^2(y^2 - 1)$ and the equilibrium points are $y_1 = -1$, $y_2 = 0$ and $y_3 = 1$. Once we are completely sure that we have all points where $f(y)$ is zero, we can find the direction of the arrows by taking test points in the open intervals obtained by removing the equilibrium points from the real line. In this case we have the intervals $(-\infty, -1)$, $(-1, 0)$, $(0, 1)$ and $(1, \infty)$. We can pick as test points -2 from the interval $(-\infty, -1)$, -0.5 form the interval $(-1, 0)$, 0.5 from the interval $(0, 1)$ and 2 from the interval $(1, \infty)$. Since $f(-2) = 12 > 0$, then the arrow in this intervals goes up. Since $f(-0.5) = f(0.5) = -0.1875 < 0$ then the arrows on the intervals $(-1, 0)$ and $(0, 1)$ go down. Finally, since $f(2) = 12 > 0$ then the arrow in the interval $(1, \infty)$ goes up. Figure (5.1.1) shows this phase line.*

Notice that if y_1 is a value in an interval with an arrow going up, then a solution of the differential equation satisfying $y(t_1) = y_1$ is increasing for values of t near t_1. Likewise, if y_1 is a value in an interval with an arrow going down, then a solution of the differential equation satisfying $y(t_1) = y_1$ is decreasing for values of t near t_1. The following theorem explains how we can do a sketch of a solution of the differential equation using the phase line.

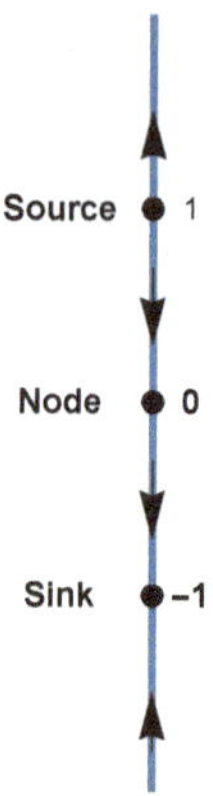

Figure 5.1.1: Phase line of the differential equation $\frac{dy}{dt} = y^2(y^2 - 1)$

Theorem 6 *If $f(y)$ is a smooth function defined for all real numbers that only vanishes at the points $y_1 < \cdots < y_n$, then the solution $y(t)$ of the initial value problem $\frac{dy}{dt} = f(y)$, $y(t_0) = y_0$ satisfies*

1. *If y_0 is in the set $\{y_1, \ldots, y_n\}$, then y_0 is an equilibrium point and the graph of $y(t)$ is a horizontal line.*

2. *If y_0 is in the interval (y_i, y_{i+1}), then the lines $y = y_i$ and $y = y_{i+1}$ are horizontal asymptotes; the function is always increasing if $f(y_0) > 0$ or always decreasing if $f(y_0) < 0$ and moreover $y(t)$ is defined for all t.*

3. *If $y_0 < y_1$, then the line $y = y_1$ is a horizontal asymptote and the solution $y(t)$ is unbounded.*

4. *If $y_n < y_0$, then the line $y = y_n$ is a horizontal asymptote and the solution $y(t)$ is unbounded.*

Example 60 *As an example of the theorem we sketch the 5 solutions of the differential equation $\frac{dy}{dt} = y^2(y^2 - 1)$ that satisfy the following initial conditions $y(1) = 0.5$, $y(0) = -0.5$, $y(1) = 1.2$, $y(-1) = -1.2$ and $y(-2) = 0.5$. Figure (5.1.2) shows these 5 graphs. Notice that the initial conditions have been highlighted with a point.*

In general it is not difficult to show that the graphs of two different solutions with values in the same interval differ by a horizontal translation.

5.2   Homework

For each of the following problems do the phase line, classifying the equilibrium points and sketch the solutions that satisfy the given initial conditions. The answer to each problem is two graphs similar to Figure (5.1.1) and Figure (5.1.2).

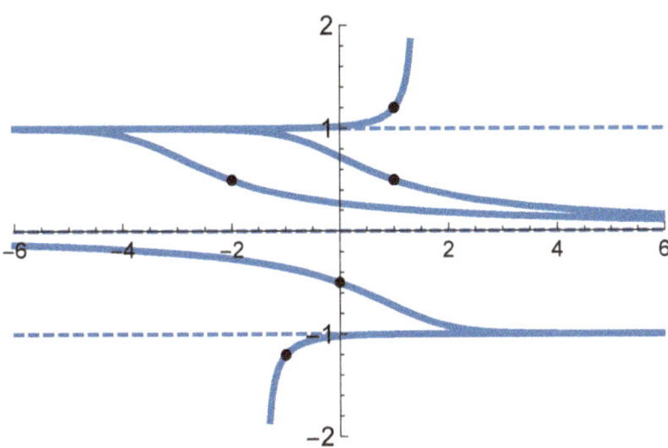

Figure 5.1.2: Phase line of the differential equation $\frac{dy}{dt} = y^2(y^2 - 1)$

1. Differential equation: $\frac{dy}{dt} = y^2 + 3y - 4$. Solutions to be considered: $y(0) = -5$, $y(-2) = 0$, $y(2) = 3$.

2. Differential equation: $\frac{dy}{dt} = y^2$. Solutions to be considered: $y(0) = -2$, $y(-2) = 0$, $y(2) = 1$.

3. Differential equation: $\frac{dy}{dt} = y$. Solutions to be considered: $y(0) = -2$, $y(-2) = 0$, $y(2) = 1$.

4. Differential equation: $\frac{dy}{dt} = y(y-1)^2$. Solutions to be considered: $y(0) = -1$, $y(-2) = 0.5$, $y(2) = 1$, $y(1) = 2$.

5. Differential equation: $\frac{dy}{dt} = y^4 + y^3 - 2y^2$. Solutions to be considered: $y(0) = 0.5$, $y(-2) = 1.5$, $y(2) = -0.5$, $y(-2) = -2$, $y(1) = -2.5$.

6. Differential equation: $\frac{dy}{dt} = (y^2 - 4)(y^2 - y - 12)$. Solutions to be considered: $y(0) = 5$, $y(-2) = 1.5$, $y(2) = -0.5$, $y(-2) = 3$, $y(1) = -4$.

Answers:

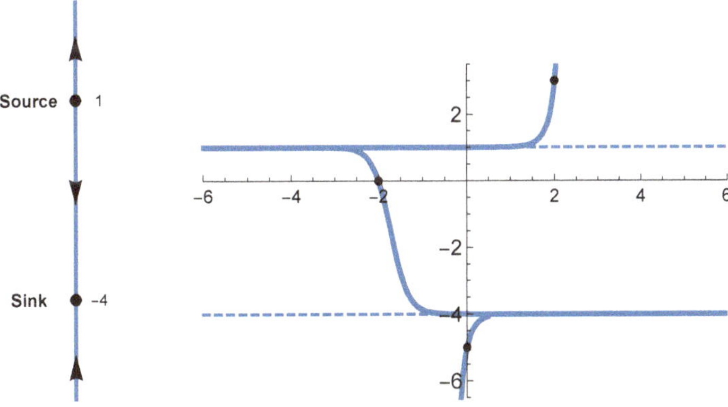

Figure 5.2.1: Answer problem 1

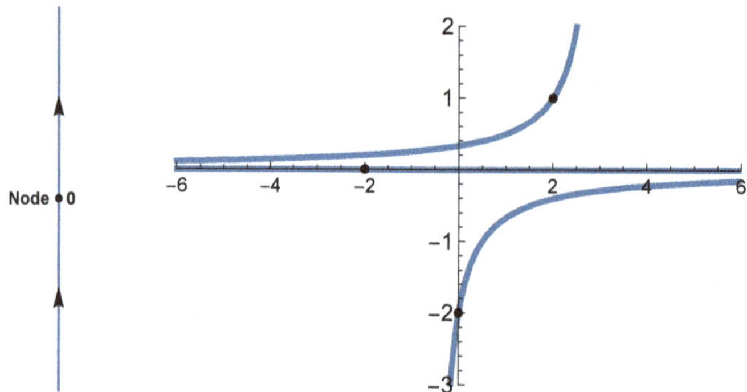

Figure 5.2.2: Answer problem 2

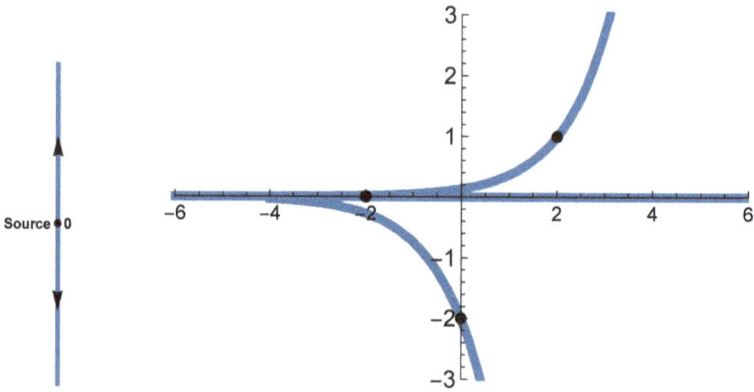

Figure 5.2.3: Answer problem 3

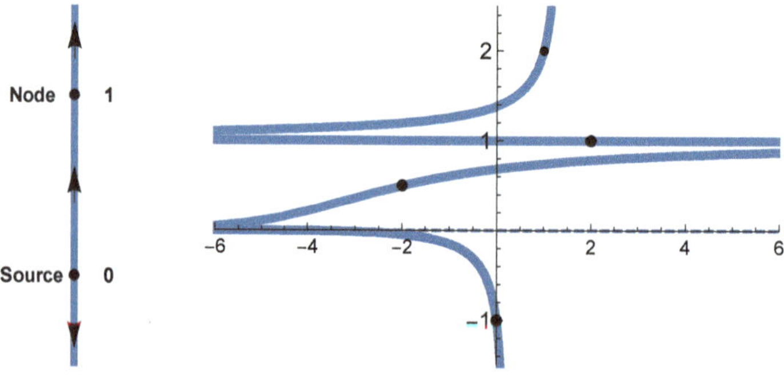

Figure 5.2.4: Answer problem 4

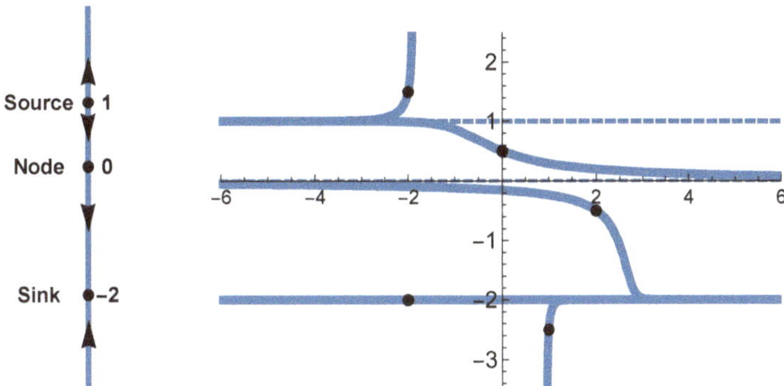

Figure 5.2.5: Answer problem 5

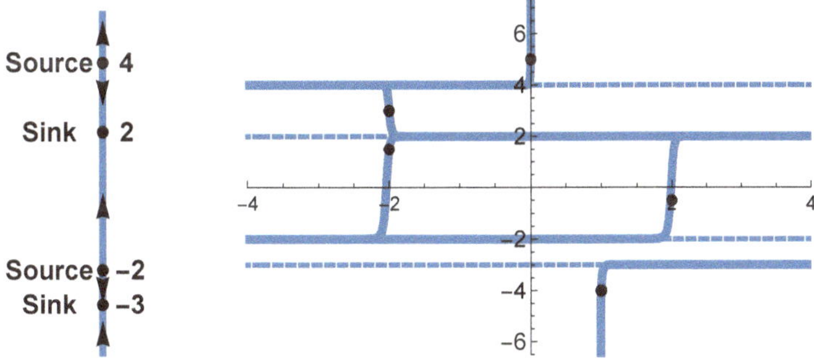

Figure 5.2.6: Answer problem 6

chapter 6

Euler Method for first order differential Equations

Numerical methods allow us to estimate solutions of differential equations. They are extremely important due to the fact that most of the differential equations cannot be solved explicitly. In this section we review the most simple of the numerical methods, the Euler method. The idea behind the method is to approximate the solution by lines over small pieces of their domain. The smaller the pieces the better the approximation will be. In this method the jumps from one value of t to another is called Δt. For example when t_0 is 2, $\Delta t = 0.01$ means that we will be considering the solutions at the values for t given by $2, 2.01, 2.02\ldots$ until we reach the last value of t that we want to consider. The main idea behind the Euler method is that if we have an initial value problem $\frac{dy}{dt} = f(y,t)$ with $y(t_0) = y_0$, then we not only know the value of y at t_0, but we also know the value of the slope of the tangent line at t_0. For example, for the initial value problem $\frac{dy}{dt} = t^2 - 2 + ty^2$ with $y(2) = 1$ we have that the graph of the solution $y(t)$ not only passes through the point $(2,1)$ but the slope of the tangent line at this point is $f(t,y) = t^2 - 2 + ty^2$ evaluated at $t = 2$ and $y = 1$, this is, it is $2^2 - 2 + 2 \times 1^2 = 4$. Figure 6.0.1 shows the solution $y(t)$ near $t = 2$.

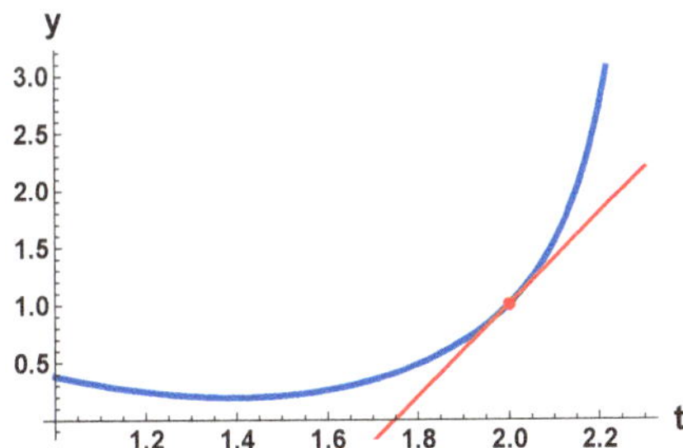

Figure 6.0.1: Solution for the system $\frac{dy}{dt} = t^2 - 2 + ty^2$ with $y(2) = 1$, near $t = 2$. Knowing the value of the function at $t = 2$ allows us to compute the exact formula for the tangent line.

The method uses the fact that near $t = 2$, the functions $y = y(t)$ and $y = 1 + 4(t - 2)$ are very close, and therefore if Δt is small, $y(2 + \Delta t)$ is close to $1 + 4\Delta t$. The next section shows the details of the method.

6.1 The method.

Let us consider the differential equation

$$\frac{dy}{dt} = f(y,t) \quad \text{with} \quad y(t_0) = y_0 \tag{6.1.1}$$

We have that the slope of the solution at t_0 is $m_0 = f(y_0, t_0)$. Therefore a linear approximation for the solution at $t_0 + \Delta t$ is $y_1 = y_0 + m_0 \Delta t$. The generalization of the previous observation lead us to the following table.

t_i	y_i	$m_i = f(y_i, t_i)$
t_0	y_0	$m_0 = f(y_0, t_0)$
$t_1 = t_0 + \Delta t$	$y_1 = y_0 + m_0 \Delta t$	$m_1 = f(y_1, t_1)$
$\vdots$	$\vdots$	$\vdots$
$t_{i+1} = t_i + \Delta t$	$y_{i+1} = y_i + m_i \Delta t$	$m_{i+1} = f(y_{i+1}, t_{i+1})$
$\vdots$	$\vdots$	$\vdots$
$t_n = t_{n-1} + \Delta t$	$y_n = y_{n-1} + m_{n-1} \Delta t$	

Notice that the values for t_0 and y_0 are taken from the initial value problem as well as the formula for the m_i. The Δt is either given or it needs to be selected according to the accuracy needs. The value y_i is the estimate given by the Euler method to the value $y(t_i)$, where $y(t)$ is the unknown solution of the initial value problem.

Example 61 *In order to find an estimate of the solution of the initial value problem $\frac{dy}{dt} = y^2 - t^2$, $y(1) = 2$ evaluated at $t = 1.1$ using the Euler method with $\Delta t = 0.02$ we do the following table*

t_i	y_i	$m_i = y_i^2 - t^2$
1	2	$m_0 = 2^2 - 1^2 = 3$
1.02	$y_1 = 2 + 3 \times 0.02 = 2.06$	$m_1 = 2.06^2 - 1.02^2 = 3.2032$
1.04	2.124064	3.430047876
1.06	2.192664958	3.684179616
1.08	2.266348550	3.969935749
1.1	2.345747265	

The answer is $\boxed{\text{the approximation for y(1.1) is 2.345747265}}$.

Example 62 *Let us use the Euler method with a differential equation that we know how to explicitly solve in order to compute the error. Let us consider the initial value problem $\frac{dy}{dt} = 2y + e^{4t}$, $y(0) = 6$ and let us find the approximation for $y(0.03)$ given by the Euler method using $\Delta t = 0.01$ and then let us find the error by solving the differential equation.*

The Euler method is given by the following table:

t_i	y_i	$m_i = 2y_i + e^{4t}$
0	6	$m_0 = 2 \times 6 + e^0 = 13$
0.01	$y_1 = 6 + 13 \times 0.01 = 6.13$	13.30081077
0.02	6.263008108	13.60930328
0.03	6.399101141	

The exact solution is of the form $y = y_H + y_p$ where $y_H = ce^{2t}$. The particular solution is of the form $y_p = Ae^{4t}$. Replacing y_p in the differential equation we obtain that $4Ae^{4t} = 2Ae^{4t} + e^{4t}$. Then, $A = 1/2$. To find c we use the initial condition, we obtain $y(0) = 6 = c + \frac{1}{2}$, this is $c = \frac{11}{2}$ and the solution is $y = \frac{11}{2}e^{2t} + \frac{1}{2}e^{4t}$. Therefore $y(0.03) = 6.403849432$ and the error is $6.403849432 - 6.399101141 = 0.004748291$.

6.2 Homework

For the following systems: (a) Solve the initial value problem. (b) Use Euler's method with $\Delta t = 0.01$ to estimate the value of the solution at $t_0 + 3\Delta t$ (c) Find the error. Round using 6 significant digits.

1. $\frac{dy}{dt} = 3y + \sin t$, $y(0) = 1$. The answer for part (a) is $y = \frac{11e^{3t}}{10} - \frac{3\sin(t)}{10} - \frac{\cos(t)}{10}$. The answer for part (b) is 1.09303. The answer for part (c) is 0.00160807.

2. $\frac{dy}{dt} = 2y + 2 + t$, $y(0) = 2$. The answer for part (a) is $y = -\frac{t}{2} + \frac{13e^{2t}}{4} - \frac{5}{4}$. The answer for part (b) is 2.18393. The answer for part (c) is 0.00204278

3. $\frac{dy}{dt} = -3y + e^{4t}$, $y(0) = 3$. The answer for part (a) is $y = \frac{20e^{-3t}}{7} + \frac{e^{4t}}{7}$. The answer for part (b) is 2.76836. The answer for part (c) is 0.003946202.

4. $\frac{dy}{dt} + \frac{1}{1+t}y = t$, $y(1) = 4$. The answer for part (a) is $y = \frac{2t^3 + 3t^2 + 43}{6t + 6}$. The answer for part (b) is 3.97075. The answer for part (c) is 0.000367758.

chapter 7

Decoupled systems

7.1 Definition

A two by two system of differential equations is called partially decoupled if one of the differential equations only depends on one function. This system can be solved by first solving the equation that only has one function and then replacing the solution in the other equation. When we solve the first equation there will be a constant in the general solution. We need to deal with this constant while solving the second equation. This constant creates a link between the two solutions. At the end the general solutions must have two constants.

Example 63 *In order to find the general solution of the system* $\begin{cases} \frac{dx}{dt} & = 3x + y \\ \frac{dy}{dt} & = 2 \end{cases}$ *we first solve the second equation because it does not depend on the function* $x(t)$. *We obtain that* $y = 2t + c_1$. *Replacing in the first equation we obtain the equation*

$$\frac{dx}{dt} = 3x + 2t + c_1$$

whose solution is $x = x_H + x_p$ *with* $x_H = c_2 e^{3t}$ *and* $x_p = At + B$ *with* A, B *satisfying*

$$A = 3(At + B) + 2t + c_1 \,,$$

this is, A and B need to satisfy the system $\begin{cases} A & = 3B + c_1 \\ 0 & = 3A + 2 \end{cases}$. *Then* $A = -\frac{2}{3}$ *and* $B = -\frac{c_1}{3} - \frac{2}{9}$. *Therefore the general solution is given by*

$$\boxed{\begin{cases} x & = c_2 e^{3t} - \frac{2}{3}t - \frac{c_1}{3} - \frac{2}{9} \\ y & = 2t + c_1 \end{cases}}$$

Example 64 *In order to find the general solution of the system* $\begin{cases} \frac{dx}{dt} & = 2x \\ \frac{dy}{dt} & = 3y + 2x^2 \end{cases}$ *we first solve the first equation because it does not depend on the function* $y(t)$. *We obtain that* $x = c_1 e^{2t}$. *Replacing in the second equation we obtain the equation*

$$\frac{dy}{dt} = 3y + 2c_1^2 \mathrm{e}^{4t}$$

whose solution is $y = y_H + y_p$ with $y_H = c_2 \mathrm{e}^{3t}$ and $y_p = A\mathrm{e}^{4t}$ with A satisfying

$$4A\mathrm{e}^{4t} = 3A\mathrm{e}^{4t} + 2c_1^2\mathrm{e}^{4t},$$

this is, $A = 2c_1^2$. Therefore the general solution is given by

$$\begin{cases} x & = c_1 \mathrm{e}^{2t} \\ y & = c_2 \mathrm{e}^{3t} + 2c_1^2 \mathrm{e}^{4t} \end{cases}$$

Example 65 *In order to find the general solution of the system* $\begin{cases} \frac{dx}{dt} & = 4x + y \\ \frac{dy}{dt} & = 3\cos(2t) \end{cases}$ *we first solve the second equation because it does not depend on the function* $x(t)$. *We obtain that* $y = \frac{3}{2}\sin(2t) + c_1$. *Replacing in the first equation we obtain the equation*

$$\frac{dx}{dt} = 4x + \frac{3}{2}\sin(2t) + c_1$$

whose solution is $x = x_H + x_p$ *with* $x_H = c_2\mathrm{e}^{4t}$ *and* $x_p = A\cos(2t) + B\sin(2t) + C$ *with A, B and C satisfying*

$$-2A\sin(2t) + 2B\cos(2t) = 4(A\cos(2t) + B\sin(2t) + C) + \frac{3}{2}\sin(2t) + c_1$$

this is,

$$\begin{cases} -2A & = 4B + \frac{3}{2} \\ 2B & = 4A \\ 0 & = 4C + c_1 \end{cases}$$

Therefore $C = -\frac{c_1}{4}$, $A = -\frac{3}{20}$ *and* $B = -\frac{3}{10}$. *The general solution is given by*

$$\begin{cases} x & = c_2\mathrm{e}^{4t} - \frac{3}{20}\cos(2t) - \frac{3}{10}\sin(2t) - \frac{c_1}{4} \\ y & = \frac{3}{2}\sin(2t) + c_1 \end{cases}$$

7.2 Homework

For the following systems: (a) Find the general solution. (b) Solve the initial value problem with $x(0) = -1$ and $y(0) = 2$ and (c) Find the point $(x(1), y(1))$ where x and y are the solution of part (b). Use 6 significant digits, no rounding

1. $\begin{cases} \frac{dx}{dt} & = 4x \\ \frac{dy}{dt} & = 2y + 2x^3 \end{cases}$. The answer for part (a) must have two constants. The answer for part (b)

   is $\left\{ x(t) = -\mathrm{e}^{4t}, y(t) = \frac{11\mathrm{e}^{2t}}{5} - \frac{\mathrm{e}^{12t}}{5} \right\}$ and the answer for part (c) is $(-54.5982, -32534.7)$

2. $\begin{cases} \frac{dx}{dt} &= -3x + y^5 \\ \frac{dy}{dt} &= 0 \end{cases}$. The answer for part (a) must have two constants. The answer for part

(b) is $\left\{ y(t) = 2, x(t) = \frac{32}{3} - \frac{35e^{-3t}}{3} \right\}$ and the answer for part (c) is $(10.0858, 2.00000)$

3. $\begin{cases} \frac{dx}{dt} &= \sin t + 2 \\ \frac{dy}{dt} &= 2y + x \end{cases}$. The answer for part (a) must have two constants. The answer for part (b)

is $\left\{ x(t) = 2t - \cos(t), y(t) = -t + \frac{21e^{2t}}{10} - \frac{\sin(t)}{5} + \frac{2\cos(t)}{5} - \frac{1}{2} \right\}$ and the answer for part
(c) is $(1.45970, 14.0648)$

4. $\begin{cases} \frac{dx}{dt} &= 3x \\ \frac{dy}{dt} &= 6y + x^2 \end{cases}$. The answer for part (a) must have two constants. The answer for part (b)

is $\left\{ x(t) = -e^{3t}, y(t) = e^{6t}t + 2e^{6t} \right\}$ and the answer for part (c) is $(-20.0855, 1210.29)$

chapter 8

Euler method for 2 by 2 systems

When we have a system of differential equations

$$\begin{cases} \frac{dx}{dt} &= f(x,y,t) \\ \frac{dy}{dt} &= g(x,y,t) \end{cases}$$

we notice that if we know the values of x and y for a particular value of t_i, more precisely, if we know that $x(t_i) = x_i$ and $y(t_i) = y_i$, then we can compute the slope of the tangent lines of the graphs $x = x(t)$ and $y = y(t)$, at $t = t_i$. We have that the slope of the tangent line for the solution $x(t)$ at t_i is $m_i = f(x_i, y_i, t_i)$ and the slope of the tangent line for the solution $y(t)$ at t_i is $n_i = g(x_i, y_i, t_i)$. For example, for the system

$$\begin{cases} \frac{dx}{dt} &= x^2 - y - t \\ \frac{dy}{dt} &= t + y^2 - x \end{cases} \quad \text{with} \quad x(1) = 2 \quad \text{and} \quad y(1) = -3 \tag{8.0.1}$$

We know that the solution $x(t)$ not only satisfies that $x(1) = 2$ but the slope of the tangent line at $(1, 2)$ is $f(x, y, t) = x^2 - y - t$ evaluated at $x = 2, t = 1$ and $y = -3$, this is, the slope is $2^2 - (-3) - 1 = 6$. Likewise, the solution $y(t)$ not only satisfies that $y(1) = -3$ but the slope of the tangent line at $(1, -3)$ is $g(x, y, t) = t + y^2 - x$ evaluated at $x = 2, t = 1$ and $y = -3$, this is, the slope is $1 + (-3)^2 - 2 = 8$. Figure 8.0.1 shows the solution $x(t)$ and $y(t)$ near $t = 1$.

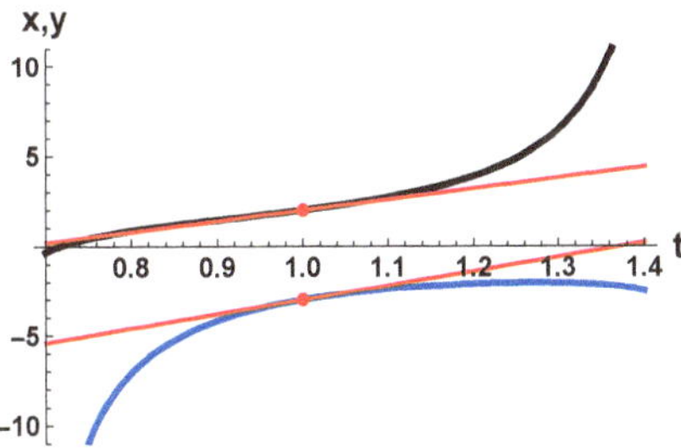

Figure 8.0.1: Solution for the system 8.1.2 near $t = 1$. Knowing the value of both functions at $t = 1$ allows us to compute the exact formula for their tangent lines

With the previous observation in mind, it is not difficult to slightly modify Euler's Method discussed in Chaper 6 to system of differential equations. The next section elaborates on the details.

8.1 The method

Let us consider the system of differential equations

$$\begin{cases} \frac{dx}{dt} = f(x,y,t) \\ \frac{dy}{dt} = g(x,y,t) \end{cases} \quad \text{with} \quad x(t_0) = x_0 \quad \text{and} \quad y(t_0) = y_0 \tag{8.1.1}$$

Following the same reasoning as the Euler method for first order differential equations, it follows that the Euler method for the system (8.1.1) is given by the following table.

t_i	x_i	y_i	$m_i = f(x_i, y_i, t_i)$	$n_i = g(x_i, y_i, t_i)$
t_0	x_0	y_0	$m_0 = f(x_0, y_0, t_0)$	$n_0 = g(x_0, y_0, t_0)$
$t_1 = t_0 + \Delta t$	$x_1 = x_0 + m_0\Delta t$	$y_1 = y_0 + n_0\Delta t$	$m_1 = f(x_1, y_1, t_1)$	$n_1 = g(x_1, y_1, t_1)$
$\vdots$	$\vdots$	$\vdots$	$\vdots$	$\vdots$
$t_{i+1} = t_i + \Delta t$	$x_{i+1} = x_i + m_i\Delta t$	$y_{i+1} = y_i + n_i\Delta t$	$m_{i+1} = f(x_{i+1}, y_{i+1}, t_{i+1})$	$n_{i+1} = f(x_{i+1}, y_{i+1}, t_{i+1})$
$\vdots$	$\vdots$	$\vdots$	$\vdots$	$\vdots$
$t_n = t_{n-1} + \Delta t$	$x_n = x_{n-1} + m_{n-1}\Delta t$	$y_n = y_{n-1} + n_{n-1}\Delta t$		

Example 66 *Let us find the approximation given by Euler's method with $\Delta t = 0.01$ to estimate at $t = 0.03$ the solution of the initial value problem*

$$\begin{cases} \frac{dx}{dt} = x^2 - y - t \\ \frac{dy}{dt} = t + y^2 - x \end{cases} \quad \text{with} \quad x(1) = 2 \quad \text{and} \quad y(1) = -3 \tag{8.1.2}$$

Notice that $t_0 = 1$, $x_0 = 2$, $y_0 = -3$, $m_i = x_i^2 - y_i - t_i$ and $n_i = t_i + y_i^2 - x_i$. Therefore,

t_i	x_i	y_i	m_i	n_i
1	2	-3	6	8
1.01	2.06	-2.92	6.1536	7.4764
1.02	2.121536	-2.845236	6.326150999	6.993831896
1.03	2.18479751	-2.775297681		

Therefore the approximation given by the Euler Method for $x(1.03)$ is 2.18479751 and for $y(1.03)$ is -2.775297681.

8.2 Changing a second order differential equation into a system of equations

In this section we explain how we can use Euler's Method to estimate the solution of second order differential equations. Let us consider the following initial value problem

$$\frac{d^2y}{dt^2} = f(\frac{dy}{dt}, y, t), \quad y(t_0) = y_0, \quad \frac{dy}{dt}(t_0) = y_0' \tag{8.2.1}$$

In the expression above, $f(\frac{dy}{dt}, y, t)$ is an expression that depends on t, y and $\frac{dy}{dt}$. The initial value problem (8.2.1) can be changed into a first order system by creating a new function $x(t)$ and making it equal to $\frac{dy}{dt}$.

$$\begin{cases} \frac{dx}{dt} &= f(x, y, t), \quad x(t_0) = y_0' \\ \frac{dy}{dt} &= x, \quad y(t_0) = y_0 \end{cases} \tag{8.2.2}$$

Let us go through an example.

Example 67 *The initial value problem $\frac{d^2y}{dt^2} + t(\frac{dy}{dt})^2 = 3y^2 - t$, $y(1) = 2$, $\frac{dy}{dt}(1) = -3$ can be written as $\frac{d^2y}{dt^2} = 3y^2 - t - t(\frac{dy}{dt})^2 = f(\frac{dy}{dt}, y, t)$. Since $x = \frac{dy}{dt}$, then*

$$\frac{dx}{dt} = \frac{d^2y}{dt^2}$$

and $x(1) = \frac{dy}{dt}(1) = -3$. Therefore we can solve the second order differential equation by solving the system

$$\begin{cases} \frac{dx}{dt} &= 3y^2 - t - tx^2, \quad x(1) = -3 \\ \frac{dy}{dt} &= x, \quad y(1) = 2 \end{cases} \tag{8.2.3}$$

We can estimate the values for $y(1.03)$, by applying the Euler's method to the system (8.2.3). If we use $\Delta t = 0.01$ we obtain that,

t_i	x_i	y_i	m_i	n_i
1	-3	2	2	-3
1.01	-2.98	1.97	1.663496	-2.98
1.02	-2.96336504	1.9402	1.315965112	-2.96336504
1.03	-2.950205389	1.91056635		

It follows that the approximation for $y(1.03)$ given by Euler's method is 1.91056635. The value -2.950205389 is the approximation for $y'(1.03)$ provided by Euler's method

Example 68 *Using the techniques we learned in the previous section we can deduce that the solution of the initial value problem $\frac{d^2y}{dt^2} + 2\frac{dy}{dt} + 2y = 10e^{2t}$, $y(0) = 2$, $\frac{dy}{dt}(0) = 1$ is $y(t) = e^{2t} + e^{-t}\cos(t)$. Therefore we have that $y(0.03) = 2.031845412$. Let us check the error of Euler's method when we use it to estimate $y(0.03)$ with $\Delta t = 0.01$. The first order system associated with the second order differential equation is*

$$\begin{cases} \frac{dx}{dt} &= 10e^{2t} - 2x - 2y, \quad x(0) = 1 \\ \frac{dy}{dt} &= x, \quad y(0) = 2 \end{cases} \tag{8.2.4}$$

We can estimate the values for $y(0.03)$. By applying Euler's method with $\Delta t = 0.01$ we get,

t_i	x_i	y_i	m_i	n_i
0	1	2	4	1
0.01	1.04	2.01	4.1020134	1.04
0.02	1.081020134	2.0204	4.205267474	1.081020134
0.03	1.123072809	2.031210201		

It follows that the approximation for $y(0.03)$ is 2.031210201. Therefore the error is $|2.031845412 - 2.031210201| = 0.000635211$

8.3 Homework

1. Use the Euler method with $\Delta t = 0.01$ to estimate at $t = 1.03$, the solution of the system

$$\begin{cases} \frac{dx}{dt} = xy - x + 6t \\ \frac{dy}{dt} = x + t - y^2 \end{cases} \quad \text{with} \quad x(1) = 1 \quad \text{and} \quad y(1) = -2$$

Answer: $x(1.03) \approx 1.088481102$, $y(1.03) = -2.061241545$.

2. Use the Euler method with $\Delta t = 0.01$ to estimate at $t = -1.97$, the solution of the system

$$\begin{cases} \frac{dx}{dt} = 6t^2 - x + y \\ \frac{dy}{dt} = t + x^2 + y + 2 \end{cases} \quad \text{with} \quad x(-2) = 3 \quad \text{and} \quad y(-2) = 0$$

Answer: $x(-1.97) \approx 3.619406350$, $y(-1.97) = 0.3128997096$.

3. Change the initial value problem into a first order system. $\frac{d^2y}{dt^2} + 2y^2 \frac{dy}{dt} - 3 \sin y = 2t^2$, $y(2) = 3$, $y'(2) = -2$

Answer: $\begin{cases} \frac{dx}{dt} = 2t^2 + 3 \sin y - 2y^2 x \\ \frac{dy}{dt} = x \end{cases} \quad \text{with} \quad x(2) = -2 \quad \text{and} \quad y(2) = 3$

4. Solve the initial value problem and use Euler's method to estimate $y(0.03)$. Then use the exact solution to find the error.
$\frac{d^2y}{dt^2} + 2\frac{dy}{dt} - 3y = 6t$, $y(0) = 3$, $\frac{dy}{dt}(0) = -6$. Answer: (a) $y(t) = -2t + \frac{25e^{-3t}}{12} + \frac{9e^t}{4} - \frac{4}{3}$, (b) 2.826246 and the error is 0.002966671.

5. Solve the initial value problem and use Euler's method to estimate $y(0.03)$. Then use the exact solution to find the error.
$\frac{d^2y}{dt^2} - 5\frac{dy}{dt} + 6y = 52 \sin(3t)$, $y(0) = 2$, $\frac{dy}{dt}(0) = 0$. Answer: (a) $y(t) = -6e^{2t} + \frac{14e^{3t}}{3} - \frac{2}{3} \sin(3t) + \frac{10}{3} \cos(3t)$, (b) 1.996495977 and the error is 0.001445188.

6. Solve the initial value problem and use Euler's method to estimate $y(0.03)$. Then use the exact solution to find the error.
$\frac{d^2y}{dt^2} - 6\frac{dy}{dt} + 9y = 2e^{3t}$, $y(0) = -1$, $\frac{dy}{dt}(0) = 3$. Answer: (a) $y(t) = e^{3t}t^2 + 6e^{3t}t - e^{3t}$, (b) -0.9011469091 and the error is 0.0049087533.

7. $\frac{d^2y}{dt^2} + 4y = 4$, $y(0) = 1$, $\frac{dy}{dt}(0) = -2$. Answer: (a) $y(t) = 1 - \sin(2t)$, (b) 0.940008 and the error is 0.0000279935.

chapter 9

Solving linear systems

In this section we give the procedures to solve/find the solution of any linear system of the form

$$\begin{cases} \frac{dx}{dt} &= ax + by \\ \frac{dy}{dt} &= cx + dy \end{cases}$$

9.1 Eigenvalues and eigenvectors of a 2 by 2 matrix

The first goal of this section is to explain how to compute the eigenvalues and eigenvectors of a 2×2 matrix.

Definition 10 *For the matrix $A = \begin{pmatrix} a & b \\ c & d \end{pmatrix}$ we define the **trace** of A and the **determinant** of A as*

$$\mathrm{Tr}(A) = a + d \quad and \quad \mathrm{Det}(A) = ad - bc$$

*The **characteristic equation** of A is the equation,*

$$\lambda^2 - \mathrm{Tr}(A)\lambda + \mathrm{Det}(A) = 0$$

Example 69 *If $A = \begin{pmatrix} 3 & 2 \\ -2 & 4 \end{pmatrix}$, then $\mathrm{Tr}(A) = 3 + 4 = 7$, $\mathrm{Det}(A) = 3 \times 4 - 2 \times (-2) = 16$ and the characteristic equation is $\lambda^2 - 7\lambda + 16 = 0$.*

Definition 11 *The product of the matrix $A = \begin{pmatrix} a & b \\ c & d \end{pmatrix}$ with the vector $v = \begin{pmatrix} v_1 \\ v_2 \end{pmatrix}$ is the vector*

$$Av = \begin{pmatrix} av_1 + bv_2 \\ cv_1 + dv_2 \end{pmatrix}$$

*We say that a vector $v \neq \begin{pmatrix} 0 \\ 0 \end{pmatrix}$ is an **eigenvector** of A if $Av = \lambda v$ for some complex number λ. The number λ is called an **eigenvalue** of A.*

Example 70 *Let A be the matrix $A = \begin{pmatrix} 4 & 1 \\ -5 & -2 \end{pmatrix}$. The vector $v = \begin{pmatrix} 1 \\ 2 \end{pmatrix}$ is not an eigenvector of A because*

$$Av = \begin{pmatrix} 4 & 1 \\ -5 & -2 \end{pmatrix} \begin{pmatrix} 1 \\ 2 \end{pmatrix} = \begin{pmatrix} 6 \\ -9 \end{pmatrix}$$

is not a multiple of $\begin{pmatrix} 1 \\ 2 \end{pmatrix}$. On the other hand, the vector $w = \begin{pmatrix} 1 \\ -1 \end{pmatrix}$ is an eigenvector of A because

$$Aw = \begin{pmatrix} 4 & 1 \\ -5 & -2 \end{pmatrix} \begin{pmatrix} 1 \\ -1 \end{pmatrix} = \begin{pmatrix} 3 \\ -3 \end{pmatrix} = 3 \begin{pmatrix} 1 \\ -1 \end{pmatrix} = 3w$$

Since $Aw = 3w$, then 3 is an eigenvalue of A. Notice that the vector $u = \begin{pmatrix} 8 \\ -8 \end{pmatrix}$ is also an eigenvector of A since we can check that $Au = 3u$. It is easy to check that an eigenvector multiplied by any nonzero number is again an eigenvector.

Example 71 *The vector $v = \begin{pmatrix} 1 \\ i \end{pmatrix}$ is an eigenvector of the matrix $\begin{pmatrix} 0 & 1 \\ -1 & 0 \end{pmatrix}$ because*

$$Av = \begin{pmatrix} 0 & 1 \\ -1 & 0 \end{pmatrix} \begin{pmatrix} 1 \\ i \end{pmatrix} = \begin{pmatrix} i \\ -1 \end{pmatrix} = i \begin{pmatrix} 1 \\ i \end{pmatrix} = iv$$

Since $Av = iv$ we conclude that i is an eigenvalue of A.

Theorem 7 *A complex number or a real number λ is an eigenvalue of $A = \begin{pmatrix} a & b \\ c & d \end{pmatrix}$ if and only if λ satisfies the quadratic equation*

$$\lambda^2 - \mathrm{Tr}(A)\lambda + \mathrm{Det}(A) = 0$$

Example 72 *Let us compute the eigenvalues of $A = \begin{pmatrix} 4 & 1 \\ -5 & -2 \end{pmatrix}$. Since $\mathrm{Tr}(A) = 2$ and $\mathrm{Det}(A) = -3$, then, the characteristic equation of A is $\lambda^2 - 2\lambda - 3 = 0$ and the eigenvalues of A are $\boxed{\lambda_1 = -1}$ and $\boxed{\lambda_2 = 3}$*

Example 73 *Let us compute the eigenvalues of $A = \begin{pmatrix} 8 & 3 \\ -12 & -4 \end{pmatrix}$. Since $\mathrm{Tr}(A) = 4$ and $\mathrm{Det}(A) = 4$, then, the characteristic equation of A is $\lambda^2 - 4\lambda + 4 = 0$ and A has only one eigenvalue. We write $\boxed{\lambda_1 = \lambda_2 = 2}$*

Example 74 *Let us compute the eigenvalues of $A = \begin{pmatrix} 2 & 2 \\ -2 & 2 \end{pmatrix}$. Since $\mathrm{Tr}(A) = 4$ and $\mathrm{Det}(A) = 8$, then, the characteristic equation of A is $\lambda^2 - 4\lambda + 8 = 0$ and the eigenvalues of A are $\boxed{\lambda_1 = 2 + 2i}$ and $\boxed{\lambda_2 = 2 - 2i}$*

The following theorem tells us how to compute the eigenvectors

Theorem 8 *If* λ *is an eigenvalue of the matrix* $A = \begin{pmatrix} a & b \\ c & d \end{pmatrix}$, *then the system*

$$\begin{cases} (a - \lambda)x + by & = & 0 \\ cx + (d - \lambda)y & = & 0 \end{cases}$$ *has infinitely many solutions. Moreover any solution* $v = \begin{pmatrix} x \\ y \end{pmatrix}$

satisfies that $Av = \lambda v$, *and therefore* $v = \begin{pmatrix} x \\ y \end{pmatrix}$ *is an eigenvector as long as* $v \neq \begin{pmatrix} 0 \\ 0 \end{pmatrix}$

Example 75 *The matrix in Example 72 has* 3 *as an eigenvalue. To compute an eigenvector associated with this eigenvalue we solve the system*

$$\begin{cases} (4 - 3)x + 1y & = & 0 \\ -5x + (-2 - 3)y & = & 0 \end{cases} \quad \text{or equivalently} \quad \begin{cases} x + y & = & 0 \\ -5x - 5y & = & 0 \end{cases}$$

Therefore $\boxed{v = \begin{pmatrix} 1 \\ -1 \end{pmatrix}}$ *is an eigenvector that satisfies* $Av = 3v$. *Notice that we could have*

chosen $v = \begin{pmatrix} -\sqrt{2} \\ \sqrt{2} \end{pmatrix}$ *as an eigenvector.*

Example 76 *The matrix on Example 74 has* $2 + 2i$ *as an eigenvalue. To compute an eigenvector associated with this eigenvalue we solve the system*

$$\begin{cases} (2 - (2 + 2i))x + 2y & = & 0 \\ -2x + (2 - (2 + 2i))y & = & 0 \end{cases} \quad \text{or equivalently} \quad \begin{cases} -2ix + 2y & = & 0 \\ -2x - 2iy & = & 0 \end{cases}$$

Therefore $\boxed{v = \begin{pmatrix} 1 \\ i \end{pmatrix}}$ *is an eigenvector that satisfies* $Av = (2 + 2i)v$.

9.2 Case I: A has two different real eigenvalues

Theorem 9 *If the matrix* $A = \begin{pmatrix} a & b \\ c & d \end{pmatrix}$ *has two different real eigenvalues* λ_1 *and* λ_2 *and* v *and* w *are eigenvectors with* $Av = \lambda_1 v$ *and* $Aw = \lambda_2 w$, *then the general solution* $x(t)$ *and* $y(t)$ *of the system*

$$\begin{cases} \frac{dx}{dt} & = & ax + by \\ \frac{dy}{dt} & = & cx + dy \end{cases}$$

is given by

$$\begin{pmatrix} x(t) \\ y(t) \end{pmatrix} = c_1 e^{\lambda_1 t} v + c_2 e^{\lambda_2 t} w$$

Example 77 *From examples 72 and 75 we have that the matrix $A = \begin{pmatrix} 4 & 1 \\ -5 & -2 \end{pmatrix}$ has eigenvalues 3 and -1 and the vector $v = \begin{pmatrix} 1 \\ -1 \end{pmatrix}$ satisfies $Av = 3v$. Using the theorem above, we have that if we want to find the general solution of the system*

$$\begin{cases} \frac{dx}{dt} & = & 4x + y \\ \frac{dy}{dt} & = & -5x - 2y \end{cases}$$

then we need to find an eigenvector w associated with the eigenvalue $\lambda = -1$. Using Theorem 8 we have that this vector w can be found from the linear system

$$\begin{cases} (4+1)x + 1y & = & 0 \\ -5x + (-2+1)y & = & 0 \end{cases} \quad \text{or equivalently} \quad \begin{cases} 5x + y & = & 0 \\ -5x - y & = & 0 \end{cases}$$

Clearly $w = \begin{pmatrix} 1 \\ -5 \end{pmatrix}$ is a solution. Therefore using Theorem 9 we obtain that

$$\begin{pmatrix} x(t) \\ y(t) \end{pmatrix} = c_1 e^{3t} \begin{pmatrix} 1 \\ -1 \end{pmatrix} + c_2 e^{-t} \begin{pmatrix} 1 \\ -5 \end{pmatrix}$$

or equivalently

$$\boxed{x = c_1 e^{3t} + c_2 e^{-t} \quad \text{and} \quad y = -c_1 e^{3t} - 5c_2 e^{-t}}$$

9.3 Case II: A has two different non-real eigenvalues

Before starting this section it is a good idea to remember that for any $\lambda = \alpha + \beta i$ we have that $e^{\lambda t} = e^{\alpha t}(\cos(\beta t) + i\sin(\beta t))$. Also, it is a good idea to remember that for a complex expression $f(z)$, the real part $\operatorname{Re}(f(z))$ and the imaginary part $\operatorname{Im}(f(z))$ are two expressions with no complex numbers in them, such that

$$f(z) = \operatorname{Re}(f(z)) + i\operatorname{Im}(f(z))$$

In other words, to find the real and imaginary parts of an expression $f(z)$ we need to expand the expression and collect all the terms that have an i. The coefficient of i is the imaginary part and the other terms that do not have i constitute the real part.

> **Theorem 10** *If the matrix* $A = \begin{pmatrix} a & b \\ c & d \end{pmatrix}$ *has two different eigenvalues* $\lambda_1 = \alpha + \beta i$ *and*
> $\lambda_2 = \alpha - \beta i$ *with* $\beta > 0$, *and* v *is an eigenvector with* $Av = \lambda_1 v$, *then the general solution*
> $x(t)$ *and* $y(t)$ *of the system*
>
> $$\begin{cases} \frac{dx}{dt} &= ax + by \\ \frac{dy}{dt} &= cx + dy \end{cases}$$
>
> *is given by* $\begin{pmatrix} x(t) \\ y(t) \end{pmatrix} =$
>
> $$c_1 e^{\alpha t} \operatorname{Re}[(\cos(\beta t) + i \sin(\beta t))\, v] + c_2 e^{\alpha t} \operatorname{Im}[(\cos(\beta t) + i \sin(\beta t))\, v]$$

Example 78 *From examples 74 and 76 we have that* $2 + 2i$ *is an eigenvalue of the matrix* $A = \begin{pmatrix} 2 & 2 \\ -2 & 2 \end{pmatrix}$ *and the vector* $v = \begin{pmatrix} 1 \\ i \end{pmatrix}$ *satisfies* $Av = (2 + 2i)v$. *Using the theorem above, we have that, in order to find the general solution of the system*

$$\begin{cases} \frac{dx}{dt} &= 2x + 2y \\ \frac{dy}{dt} &= -2x + 2y \end{cases}$$

we need to find the real and imaginary parts of $e^{(2+2i)t}v$. *Let us do this,*

$$
\begin{aligned}
e^{(2+2i)t}v &= e^{2t}(\cos(2t) + i\sin(2t)) \begin{pmatrix} 1 \\ i \end{pmatrix} = \begin{pmatrix} e^{2t}(\cos(2t) + i\sin(2t)) \\ e^{2t}(i\cos(2t) + i^2\sin(2t)) \end{pmatrix} \\
&= \begin{pmatrix} e^{2t}(\cos(2t) + i\sin(2t)) \\ e^{2t}(-\sin(2t) + i\cos(2t)) \end{pmatrix}
\end{aligned}
$$

Therefore, $\operatorname{Re}[e^{(2+2i)t}v] = \begin{pmatrix} e^{2t}\cos(2t) \\ -e^{2t}\sin(2t) \end{pmatrix}$ *and* $\operatorname{Im}[e^{(2+2i)t}v] = \begin{pmatrix} e^{2t}\sin(2t) \\ e^{2t}\cos(2t) \end{pmatrix}$ *and the general solution of the system is*

$$\begin{pmatrix} x(t) \\ y(t) \end{pmatrix} = c_1 \begin{pmatrix} e^{2t}\cos(2t) \\ -e^{2t}\sin(2t) \end{pmatrix} + c_2 \begin{pmatrix} e^{2t}\sin(2t) \\ e^{2t}\cos(2t) \end{pmatrix}$$

or equivalently,

$$\boxed{x(t) = c_1 e^{2t}\cos(2t) + c_2 e^{2t}\sin(2t) \quad and \quad y(t) = -c_1 e^{2t}\sin(2t) + c_2 e^{2t}\cos(2t)}$$

9.4 Case III: A has only one eigenvalue

This case is easier than the previous two but different. Here there are two reasons why this case is different. First, we do not need to find eigenvectors and, second, if we are presented with an initial value problem with initial conditions at $t = 0$, then, we do not need to find the general solution first, the method allows us to find the solution of this initial value problem right away.

Theorem 11 *If the matrix $A = \begin{pmatrix} a & b \\ c & d \end{pmatrix}$ has only one eigenvalue $\lambda_1 = \lambda_2 = \lambda$ then, the solution of the initial value problem*

$$\begin{cases} \frac{dx}{dt} &= ax + by \\ \frac{dy}{dt} &= cx + dy \end{cases} \quad with \quad x(0) = x_0 \quad y(0) = y_0$$

is

$$\begin{pmatrix} x(t) \\ y(t) \end{pmatrix} = e^{\lambda t} V_0 + t e^{\lambda t} V_1$$

where,

$$V_0 = \begin{pmatrix} x_0 \\ y_0 \end{pmatrix} \quad and \quad V_1 = \begin{pmatrix} a - \lambda & b \\ c & d - \lambda \end{pmatrix} V_0$$

If we want to find the general solution we just change $V_0 = \begin{pmatrix} x_0 \\ y_0 \end{pmatrix}$ by $V_0 = \begin{pmatrix} c_1 \\ c_2 \end{pmatrix}$

Example 79 *From example 73 we have that 2 is the only eigenvalue of the the matrix $A = \begin{pmatrix} 8 & 3 \\ -12 & -4 \end{pmatrix}$. Using the theorem above, we have that if we want to find the general solution of the system*

$$\begin{cases} \frac{dx}{dt} &= 8x + 3y \\ \frac{dy}{dt} &= -12x - 4y \end{cases}$$

then we just compute the vectors $V_0 = \begin{pmatrix} c_1 \\ c_2 \end{pmatrix}$ and

$$V_1 = \begin{pmatrix} 8 - 2 & 3 \\ -12 & -4 - 2 \end{pmatrix} \begin{pmatrix} c_1 \\ c_2 \end{pmatrix} = \begin{pmatrix} 6 & 3 \\ -12 & -6 \end{pmatrix} \begin{pmatrix} c_1 \\ c_2 \end{pmatrix} = \begin{pmatrix} 6c_1 + 3c_2 \\ -12c_1 - 6c_2 \end{pmatrix}$$

The general solution is

$$\begin{pmatrix} x(t) \\ y(t) \end{pmatrix} = e^{2t} \begin{pmatrix} c_1 \\ c_2 \end{pmatrix} + t e^{2t} \begin{pmatrix} 6c_1 + 3c_2 \\ -12c_1 - 6c_2 \end{pmatrix}$$

or equivalently,

$$\boxed{x(t) = c_1 e^{2t} + (6c_1 + 3c_2) t e^{2t} \quad and \quad y(t) = c_2 e^{2t} - (12c_1 + 6c_2) t e^{2t}}$$

Example 80 *Let us solve the following initial value problem*

$$\begin{cases} \frac{dx}{dt} &= x + 8y \\ \frac{dy}{dt} &= -2x - 7y \end{cases} \quad with \quad x(0) = 2, \ y(0) = -3$$

We first compute the matrix of the system. In this case we have that $A = \begin{pmatrix} 1 & 8 \\ -2 & -7 \end{pmatrix}$. We have that $\mathrm{Tr}(A) = -6$, $\mathrm{Det}(A) = 9$ and the characteristic equation of A is $\lambda^2 + 6\lambda + 9 = 0$. This time *the only solution of the characteristic equation is $\lambda_1 = \lambda_2 = -3$. Since -3 is the only eigenvalue of A, then we can use Theorem 11 to find the solution. We have that $V_0 = \begin{pmatrix} 2 \\ -3 \end{pmatrix}$ and*

$$V_1 = \begin{pmatrix} 1+3 & 8 \\ -2 & -7+3 \end{pmatrix} \begin{pmatrix} 2 \\ -1 \end{pmatrix} = \begin{pmatrix} 4 & 8 \\ -2 & -4 \end{pmatrix} \begin{pmatrix} 2 \\ -3 \end{pmatrix} = \begin{pmatrix} -16 \\ 8 \end{pmatrix}$$

Therefore, the solution of the initial value problem is

$$\begin{pmatrix} x(t) \\ y(t) \end{pmatrix} = e^{-3t} \begin{pmatrix} 2 \\ -3 \end{pmatrix} + te^{-3t} \begin{pmatrix} -16 \\ 8 \end{pmatrix}$$

or equivalently,

$$\boxed{x(t) = 2e^{-3t} - 16te^{-3t} \quad and \quad y(t) = -3e^{-3t} + 8te^{-3t}}$$

9.5 More examples

In this section we present more examples of the cases.

Example 81 *Let us find the general solution of the system*

$$\begin{cases} \frac{dx}{dt} &= 2x + 3y \\ \frac{dy}{dt} &= 2y \end{cases}$$

We first compute the matrix of the system. In this case we have that $A = \begin{pmatrix} 2 & 3 \\ 0 & 2 \end{pmatrix}$. We have that $\mathrm{Tr}(A) = 4$, $\mathrm{Det}(A) = 4$ and the characteristic equation of A is $\lambda^2 - 4\lambda + 4 = 0$. The only solution of the characteristic equation is $\lambda_1 = \lambda_2 = 2$. Since 2 is the only eigenvalue of A, then we can use Theorem 11 to find the solution. Since we are looking for the general solution, we take $V_0 = \begin{pmatrix} c_1 \\ c_2 \end{pmatrix}$ and,

$$V_1 = \begin{pmatrix} 2-2 & 3 \\ 0 & 2-2 \end{pmatrix} \begin{pmatrix} c_1 \\ c_2 \end{pmatrix} = \begin{pmatrix} 0 & 3 \\ 0 & 0 \end{pmatrix} \begin{pmatrix} c_1 \\ c_2 \end{pmatrix} = \begin{pmatrix} 3c_2 \\ 0 \end{pmatrix}$$

Therefore, the general solution is given by

$$\begin{pmatrix} x(t) \\ y(t) \end{pmatrix} = e^{2t} \begin{pmatrix} c_1 \\ c_2 \end{pmatrix} + te^{2t} \begin{pmatrix} 3c_2 \\ 0 \end{pmatrix}$$

or equivalently,

$$\boxed{x(t) = c_1 e^{2t} + 3c_2 te^{2t} \quad and \quad y(t) = c_2 e^{2t}}$$

Example 82 *Let us solve the following initial value problem*

$$\begin{cases} \frac{dx}{dt} &= 5x - 3y \\ \frac{dy}{dt} &= -5x + 7y \end{cases} \quad with \quad x(0) = 2,\ y(0) = -4$$

We first compute the matrix of the system. In this case we have that $A = \begin{pmatrix} 5 & -3 \\ -5 & 7 \end{pmatrix}$. *We have that* $\mathrm{Tr}(A) = 12$, $\mathrm{Det}(A) = 20$ *and the characteristic equation of A is* $\lambda^2 - 12\lambda + 20 = 0$. *The solutions of the characteristic equation are* $\lambda_1 = 2$ *and* $\lambda_2 = 10$.

We proceed to compute the eigenvectors using Theorem 8. To compute the eigenvector associated with the eigenvalue $\lambda_1 = 2$ *we solve the system*

$$\begin{cases} (5-2)x - 3y &= 0 \\ -5x + (7-2)y &= 0 \end{cases} \quad \text{or equivalently} \quad \begin{cases} 3x - 3y &= 0 \\ -5x + 5y &= 0 \end{cases}$$

Clearly $v = \begin{pmatrix} 1 \\ 1 \end{pmatrix}$ *is a solution and therefore v is a eigenvector of the eigenvalue* $\lambda_1 = 2$. *To compute the eigenvector associated with the eigenvalue* $\lambda_2 = 10$ *we solve the system*

$$\begin{cases} (5-10)x - 3y &= 0 \\ -5x + (7-10)y &= 0 \end{cases} \quad \text{or equivalently} \quad \begin{cases} -5x - 3y &= 0 \\ -5x - 3y &= 0 \end{cases}$$

Clearly $w = \begin{pmatrix} -3 \\ 5 \end{pmatrix}$ *is a solution and therefore w is a eigenvector of the eigenvalue* $\lambda_1 = 10$. *Therefore using Theorem 9 we obtain that the general solution is given by*

$$\begin{pmatrix} x(t) \\ y(t) \end{pmatrix} = c_1 e^{2t} \begin{pmatrix} 1 \\ 1 \end{pmatrix} + c_2 e^{10t} \begin{pmatrix} -3 \\ 5 \end{pmatrix}$$

or equivalently,

$$x(t) = c_1 e^{2t} - 3c_2 e^{10t} \quad \text{and} \quad y(t) = c_1 e^{2t} + 5c_2 e^{10t}$$

Using the initial conditions we find that c_1 and c_2 need to satisfy the following conditions.

$$c_1 - 3c_2 = 2 \quad \text{and} \quad c_1 + 5c_2 = -4$$

which give us that $c_1 = -\frac{1}{4}$, and $c_2 = -\frac{3}{4}$. Therefore the solution of the initial value problem is

$$\boxed{x(t) = -\frac{1}{4} e^{2t} + \frac{9}{4} e^{10t} \quad \text{and} \quad y(t) = -\frac{1}{4} e^{2t} - \frac{15}{4} e^{10t}}$$

Example 83 *Let us solve the following initial value problem*

$$\begin{cases} \frac{dx}{dt} &= 2y \\ \frac{dy}{dt} &= -2x \end{cases} \quad \text{with} \quad x(0) = 2,\ y(0) = -1$$

We first compute the matrix of the system. In this case we have that $A = \begin{pmatrix} 0 & 2 \\ -2 & 0 \end{pmatrix}$. *We have that* $\mathrm{Tr}(A) = 0$, $\mathrm{Det}(A) = 4$ *and the characteristic equation of A is* $\lambda^2 + 4 = 0$. *The solutions of the characteristic equation are* $\lambda_1 = 2i$ *and* $\lambda_2 = -2i$. *In this case we only need to compute the eigenvector associated with* $\lambda_1 = 2i$. *To compute this eigenvector we solve the system*

$$\begin{cases} (0-2i)x + 2y &= 0 \\ -2x + (0-2i)y &= 0 \end{cases} \quad \text{or equivalently} \quad \begin{cases} -2ix + 2y &= 0 \\ -2x - 2iy &= 0 \end{cases}$$

the vector $v = \begin{pmatrix} 1 \\ i \end{pmatrix}$ satisfies both equations and therefore it is an eigenvector. In order to find the general solution we need to find the real and imaginary part of $e^{2it}v$. Let us do this,

$$e^{2it}v = (\cos(2t) + i\sin(2t))\begin{pmatrix} 1 \\ i \end{pmatrix} = \begin{pmatrix} \cos(2t) + i\sin(2t) \\ i\cos(2t) + i^2\sin(2t) \end{pmatrix} = \begin{pmatrix} \cos(2t) + i\sin(2t) \\ -\sin(2t) + i\cos(2t) \end{pmatrix}$$

Therefore, $\mathrm{Re}[e^{2it}v] = \begin{pmatrix} \cos(2t) \\ -\sin(2t) \end{pmatrix}$ and $\mathrm{Im}[e^{2it}v] = \begin{pmatrix} \sin(2t) \\ \cos(2t) \end{pmatrix}$ and the general solution of the system is

$$\begin{pmatrix} x(t) \\ y(t) \end{pmatrix} = c_1 \begin{pmatrix} \cos(2t) \\ -\sin(2t) \end{pmatrix} + c_2 \begin{pmatrix} \sin(2t) \\ \cos(2t) \end{pmatrix}$$

or equivalently,

$$x(t) = c_1\cos(2t) + c_2\sin(2t) \quad and \quad y(t) = -c_1\sin(2t) + c_2\cos(2t)$$

Using the initial conditions we find that c_1 and c_2 need to satisfy the following conditions $c_1 = 2$ and $c_2 = -1$. Therefore the solution of the initial value problem is

$$\boxed{x(t) = 2\cos(2t) - \sin(2t) \quad and \quad y(t) = -2\sin(2t) - \cos(2t)}$$

9.6 Homework

1. Solve the initial value problem $\begin{cases} \frac{dx}{dt} = 11x + 10y \\ \frac{dy}{dt} = -13x - 11y \end{cases}$, $x(0) = 1$, $y(0) = -1$. Answer: $x(t) = \frac{1}{3}\sin(3t) + \cos(3t)$, $y(t) = -\frac{2}{3}\sin(3t) - \cos(3t)$.

2. Solve the initial value problem $\begin{cases} \frac{dx}{dt} = 5x - 2y \\ \frac{dy}{dt} = 2x + y \end{cases}$, $x(0) = 3$, $y(0) = -2$. Answer: $x(t) = 10e^{3t}t + 3e^{3t}$, $y(t) = 10e^{3t}t - 2e^{3t}$.

3. Solve the initial value problem $\begin{cases} \frac{dx}{dt} = 2x + 2y \\ \frac{dy}{dt} = x + 3y \end{cases}$, $x(0) = 0$, $y(0) = -3$. Answer: $x(t) = 2e^t - 2e^{4t}$, $y(t) = -e^t - 2e^{4t}$.

4. Solve the initial value problem $\begin{cases} \frac{dx}{dt} = 2x + 3y \\ \frac{dy}{dt} = 3x + 2y \end{cases}$, $x(0) = 2$, $y(0) = 0$ Answer: $x(t) = e^{-t} + e^{5t}$, $y(t) = e^{5t} - e^{-t}$.

5. Solve the initial value problem $\begin{cases} \frac{dx}{dt} = 2x + 3y \\ \frac{dy}{dt} = -3x + 2y \end{cases}$, $x(0) = 2$, $y(0) = 1$. Answer: $x(t) = e^{2t}\sin(3t) + 2e^{2t}\cos(3t)$, $y(t) = e^{2t}\cos(3t) - 2e^{2t}\sin(3t)$

6. Solve the initial value problem $\begin{cases} \frac{dx}{dt} &= 2x+y \\ \frac{dy}{dt} &= -13x-2y \end{cases}$, $x(0)=4$, $y(0)=1$. Answer: $x(t)=3\sin(3t)+4\cos(3t)$, $y(t)=\cos(3t)-18\sin(3t)$.

7. Solve the initial value problem $\begin{cases} \frac{dx}{dt} &= 4x-y \\ \frac{dy}{dt} &= x+2y \end{cases}$, $x(0)=-3$, $y(0)=2$. Answer: $x(t)=-5e^{3t}t-3e^{3t}$, $y(t)=-5e^{3t}t+2e^{3t}$.

8. Solve the initial value problem. $\begin{cases} \frac{dx}{dt} &= 3x+2y \\ \frac{dy}{dt} &= -5x-3y \end{cases}$, $x(0)=1$, $y(0)=-2$. Answer: $x(t)=\cos(t)-\sin(t)$, $y(t)=\sin(t)-2\cos(t)$.

9. Solve the initial value problem. $\begin{cases} \frac{dx}{dt} &= 3x+5y \\ \frac{dy}{dt} &= -4x-5y \end{cases}$, $x(0)=1$, $y(0)=-2$. Answer: $x(t)=e^{-t}\cos(2t)-3e^{-t}\sin(2t)$, $y(t)=2e^{-t}\sin(2t)-2e^{-t}\cos(2t)$.

10. Solve the initial value problem. $\begin{cases} \frac{dx}{dt} &= 3x-2y \\ \frac{dy}{dt} &= -6x+4y \end{cases}$, $x(0)=7$, $y(0)=14$. Answer: $x(t)=8-e^{7t}$, $y(t)=2e^{7t}+12$.

11. Solve the initial value problem. $\begin{cases} \frac{dx}{dt} &= 3x-2y \\ \frac{dy}{dt} &= 2x+7y \end{cases}$, $x(0)=1$, $y(0)=-2$. Answer: $x(t)=2e^{5t}t+e^{5t}$, $y(t)=-2e^{5t}t-2e^{5t}$.

12. Solve the initial value problem. $\begin{cases} \frac{dx}{dt} &= x+4y \\ \frac{dy}{dt} &= x+y \end{cases}$, $x(0)=4$, $y(0)=-6$. Answer: $x(t)=8e^{-t}-4e^{3t}$, $y(t)=-4e^{-t}-2e^{3t}$.

13. Solve the initial value problem. $\begin{cases} \frac{dx}{dt} &= 2x+3y \\ \frac{dy}{dt} &= -5x-6y \end{cases}$, $x(0)=2$, $y(0)=-4$. Answer: $x(t)=3e^{-3t}-e^{-t}$, $y(t)=e^{-t}-5e^{-3t}$.

14. Solve the initial value problem. $\begin{cases} \frac{dx}{dt} &= -2x+y \\ \frac{dy}{dt} &= x-2y \end{cases}$, $x(0)=2$, $y(0)=-4$. Answer: $x(t)=3e^{-3t}-e^{-t}$, $y(t)=-3e^{-3t}-e^{-t}$.

15. Solve the initial value problem. $\begin{cases} \frac{dx}{dt} &= 4x+y \\ \frac{dy}{dt} &= 3x+2y \end{cases}$, $x(0)=0$, $y(0)=2$. Answer: $x(t)=\frac{e^{5t}}{2}-\frac{e^{t}}{2}$, $y(t)=\frac{3e^{t}}{2}+\frac{e^{5t}}{2}$.

16. Solve the initial value problem. $\begin{cases} \frac{dx}{dt} &= x+2y \\ \frac{dy}{dt} &= 2x+4y \end{cases}$, $x(0)=0$, $y(0)=5$. Answer: $x(t)=2e^{5t}-2$, $y(t)=4e^{5t}+1$.

chapter 10

Phase portrait of 2 by 2 linear systems

10.1   Equilibrium points and phase portrait

Definition 12 *We say that (x_0, y_0) is an equilibrium point of the system*

$$\begin{cases} \frac{dx}{dt} & = & f(t, x, y) \\ \frac{dy}{dt} & = & g(t, x, y) \end{cases}$$

If for all t, $f(t, x_0, y_0) = 0$ and $g(t, x_0, y_0) = 0$. Notice that (x_0, y_0) is an equilibrium point if and only if $\{x(t) = x_0, \quad y(t) = y_0\}$ is a solution of the system. The vector field $V(t, x, y) = (f(t, x, y), g(t, x, y))$ is called the vector field associated with the differential equation.

Example 84 *The equilibrium points of the system $\begin{cases} \frac{dx}{dt} & = & x(x + y - 2) \\ \frac{dy}{dt} & = & y(2x - y - 1) \end{cases}$ are points $(0, 0)$, $(2, 0)$, $(0, -1)$ and $(1, 1)$. The fact that these are the only equilibrium points follows from the fact that the system of equations,*

$$x(x + y - 2) = 0 \quad and \quad y(2x - y - 1) = 0$$

is equivalent to the equations

$$(x = 0 \quad and \quad y = 0) \quad or \quad (x = 0 \quad and \quad 2x - y - 1 = 0) \quad or$$

$$(x + y - 2 = 0 \quad and \quad y = 0) \quad or \quad (x + y - 2 = 0 \quad and \quad 2x - y - 1 = 0)$$

Proposition 7 *For a linear system $\begin{cases} \frac{dx}{dt} & = & ax + by \\ \frac{dy}{dt} & = & cx + dy \end{cases}$ we have that every point (x_0, y_0) is an equilibrium point if the matrix $A = \begin{pmatrix} a & b \\ c & d \end{pmatrix}$ is the zero matrix. If A is not the zero matrix and $\mathrm{Det}(A) = 0$, then the equilibrium points of the system are the points in the line $\{c(v_1, v_2) : c \in \mathbb{R}\}$ where $v = \begin{pmatrix} v_1 \\ v_2 \end{pmatrix}$ satisfies that $Av = 0v$. When $\mathrm{Det}(A) \neq 0$, then the only equilibrium point is the point $(0, 0)$.*

Definition 13 *Let us consider the autonomous system*

$$\begin{cases} \frac{dx}{dt} &= f(x,y) \\ \frac{dy}{dt} &= g(x,y) \end{cases} \tag{10.1.1}$$

For every point $p_1 = (x_1, y_1)$ we define

$$C_{p_1} = \{p = (x(t_1), y(t_1)) : (x(t), y(t)) \text{ is a solution with } (x(0), y(0)) = (x_1, y_1) \}$$

We call C_{p_1} the orbit of p_1.

Proposition 8 *Let us consider the autonomous system (10.1.1).*

- *The orbit C_{p_1} has only a point if and only if p_1 is an equilibrium point.*

- *Two different orbits C_{p_1} and C_{p_2} do not intersect, that is, they do not have any points in common.*

Definition 14 *The phase portrait of the autonomous system (10.1.1) is the union of all C_p with $p \in \mathbb{R}^2$. When an orbit C_p is not a point it needs to be considered as an oriented curve with the orientation given by the solution $(x(t), y(t))$.*

Remark 5 *For the autonomous system* $\begin{cases} \frac{dx}{dt} &= f(x,y) \\ \frac{dy}{dt} &= g(x,y) \end{cases}$ *the vector field $V(x,y) = (f(x,y), g(x,y))$ has the property that for any point $p = (x_0, y_0)$, the orbit C_p that contains p has the vector $V(x_0, y_0)$ as a tangent vector. This property will be used to decide the directions of the orbits*

10.2 Phase portrait of Linear systems

In this section we explain the phase portrait of the system

$$\begin{cases} \frac{dx}{dt} &= ax + by \\ \frac{dy}{dt} &= cx + dy \end{cases} \tag{10.2.1}$$

in terms of eigenvalues of the matrix $A = \begin{pmatrix} a & b \\ c & d \end{pmatrix}$. The following proposition shows that when A has two nonzero real eigenvalues then the phase portrait gets divided into four regions that we may think of as quadrants with angles that are not necessarily 90 degrees. The boundary of these quadrants are four semi-lines. see Figure 10.2.1.

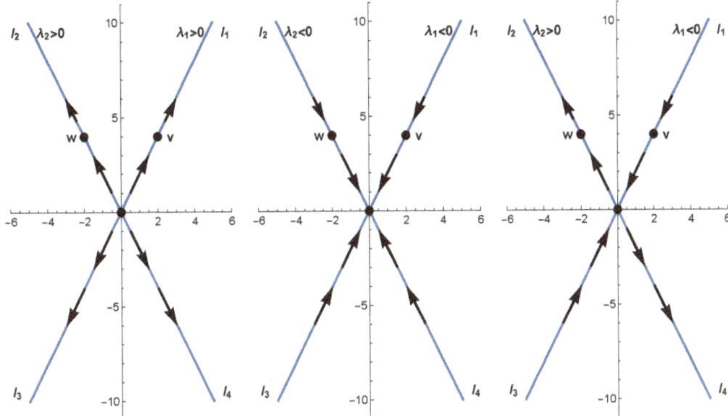

Figure 10.2.1: Semi-lines in the phase portrait.

Proposition 9 *If the matrix A has two nonzero real eigenvalues λ_1 and λ_2 and v and w are eigenvectors with $Av = \lambda_1 v$ and $Aw = \lambda_2 w$, then the semi-lines $l_1 = \{sv : s > 0\}$, $l_2 = \{sw : s > 0\}$, $l_3 = \{sv : s < 0\}$ and $l_4 = \{sw : s < 0\}$ are part of the phase portrait. If $\lambda_1 > 0$ then l_1 and l_3 go in the direction away from the origin. If $\lambda_1 < 0$ then l_1 and l_3 go in the direction toward the origin. If $\lambda_2 > 0$ then l_2 and l_4 go in the direction away from the origin. If $\lambda_2 < 0$ then l_2 and l_4 go in the direction toward the origin.*

Proof To prove that l_1 is part of the phase portrait it is enough to check that if $v = \begin{pmatrix} v_1 \\ v_2 \end{pmatrix}$ then, $\{x(t) = e^{\lambda_1 t} v_1 \quad y(t) = e^{\lambda_1 t} v_2\}$ is a solution of the system. Notice that if $\lambda_1 > 0$ then l_1 and l_3 go in the direction away from the origin because $\lim_{t \to \infty} e^{\lambda_1 t} = \infty$. Also, if $\lambda_1 < 0$ then l_1 and l_3 go in the direction toward the origin because $\lim_{t \to \infty} e^{\lambda_1 t} = 0$. The proof is similar for the lines l_2 and l_4.

10.2.1   Case 1: Two positive eigenvalues

Let us assume that the vectors v and w are two eigenvectors of A such that $Av = \lambda_1 v$ and $Aw = \lambda_2 w$ with $0 < \lambda_1 < \lambda_2$. Using Proposition 9 we obtain four semi-lines with direction going away from the origin. Also we have a dot in the origin representing the only equilibrium point of the system; the point $(0, 0)$. The general solution of the system is given by

$$\begin{pmatrix} x(t) \\ y(t) \end{pmatrix} = c_1 e^{\lambda_1 t} v + c_2 e^{\lambda_2 t} w$$

From the expression above we notice that if $c_1 > 0$ and $c_2 > 0$ then the solution lies between the semi-lines l_1 and l_2 because every $\begin{pmatrix} x(t) \\ y(t) \end{pmatrix}$ is a linear combination of the form $s_1 v + s_2 w$ with positive coefficients s_1 and s_2. We notice that when t is big and negative, then the solution is very close to the origin and since

$$c_1 e^{\lambda_1 t} v + c_2 e^{\lambda_2 t} w = e^{\lambda_1 t} \left(c_1 v + e^{(\lambda_2 - \lambda_1)t} w \right)$$

we conclude that the solution is almost a multiple of v since $\lim_{t \to -\infty} e^{(\lambda_2 - \lambda_1)t} = 0$. We can say that the dominant part is the term $c_1 v$. Likewise, when t goes to infinity we have that the dominant term is $c_2 w$ and for this reason, away from the origin, the orbits must look like a line with direction given by the vector w. For this case we say that $(0,0)$ is a **source**. Here is an example for this case.

Example 85 *Let us consider the system*

$$\begin{cases} \frac{dx}{dt} &= 8x + 2y \\ \frac{dy}{dt} &= 6x + 4y \end{cases} \tag{10.2.2}$$

In this case $A = \begin{pmatrix} 8 & 2 \\ 6 & 4 \end{pmatrix}$. A direct computation shows that the eigenvalues of A are 2 and 10. Moreover, if $v = \begin{pmatrix} -1 \\ 3 \end{pmatrix}$ and $w = \begin{pmatrix} 1 \\ 1 \end{pmatrix}$, then $Av = 2v$ and $Aw = 10w$. Figure 10.2.2 shows the phase portrait of the differential system.

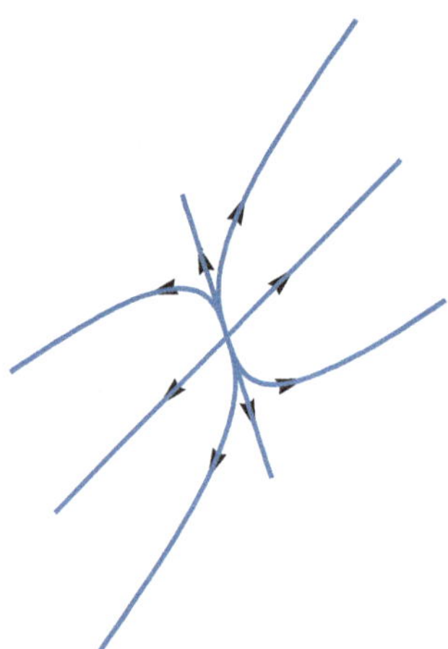

Figure 10.2.2: Phase line example with two real positive eigenvales. We can see that near the origin the orbits are tangent to the line with direction given by the vector $(-1, 3)$, which is the eigenvector associated with $\lambda = 2$. Far away from the origin the orbits are close to being lines with direction vector $(1, 1)$ which is the eigenvector associated with $\lambda = 10$.

10.2.2 Case 2: Two negative eigenvalues

Let us assume that the vectors v and w are two eigenvectors of A such that $Av = \lambda_1 v$ and $Aw = \lambda_2 w$ with $\lambda_2 < \lambda_1 < 0$. Using Proposition 9 we obtain four semi-lines with direction going toward the origin. Also we have a dot in the origin representing the only equilibrium point of the system; the point $(0,0)$. The general solution of the system is given by

$$\begin{pmatrix} x(t) \\ y(t) \end{pmatrix} = c_1 e^{\lambda_1 t} v + c_2 e^{\lambda_2 t} w$$

From the expression above we notice that if $c_1 > 0$ and $c_2 > 0$ then the solution lies between the semi-lines l_1 and l_2 because every $\begin{pmatrix} x(t) \\ y(t) \end{pmatrix}$ is a linear combination of the form $s_1 v + s_2 w$ with positive coefficients s_1 and s_2. We notice that when t is big and positive, then the solution is very close to the origin and since

$$c_1 \mathrm{e}^{\lambda_1 t} v + c_2 \mathrm{e}^{\lambda_2 t} w = \mathrm{e}^{\lambda_1 t} \left(c_1 v + \mathrm{e}^{(\lambda_2 - \lambda_1)t} w \right)$$

we conclude that the solution is almost a multiple of v since $\lim_{t \to \infty} \mathrm{e}^{(\lambda_2 - \lambda_1)t} = 0$. We can say that when $t \to \infty$, the dominant part is the term $c_1 v$. Likewise, when t goes to negative infinity we have that the dominant term is $c_2 w$ and for this reason the orbits must look like a line with direction given by the vector w. For this case we say that $(0,0)$ is a **sink**. Here is an example for this case.

Example 86 *Let us consider the system*

$$\begin{cases} \frac{dx}{dt} &= -x + 2y \\ \frac{dy}{dt} &= -3y \end{cases} \tag{10.2.3}$$

In this case $A = \begin{pmatrix} -1 & 2 \\ 0 & -3 \end{pmatrix}$. A direct computation shows that the eigenvalues of A are -1 and -3. Moreover, if $v = \begin{pmatrix} 1 \\ 0 \end{pmatrix}$ and $w = \begin{pmatrix} -1 \\ 1 \end{pmatrix}$, then $Av = -v$ and $Aw = -3w$. Figure 10.2.3 shows the phase portrait of the differential system.

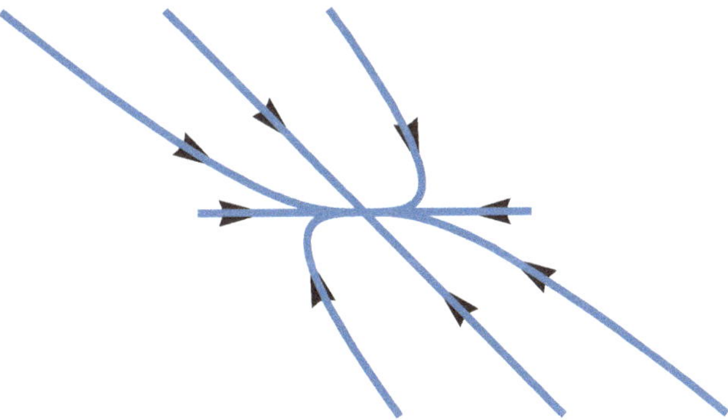

Figure 10.2.3: Phase line example with two real negative eigenvales.

The following remark is useful to decide the form of the orbits for the cases 1 and 2

Remark 6 *Near the origin, the orbits (that are not semi-lines) must be almost tangent to the direction given by the eigenvector associated with the eigenvalue that is closer to zero. Away from the origin the orbit must be almost parallel to the direction given by the eigenvector associated with the eigenvalue that is farther away from the origin. This rule works when we have two real eigenvalues with the same sign.*

10.2.3   Case 3: One negative and one positive eigenvalue

Let us assume that the vectors v and w are two eigenvectors of A such that $Av = \lambda_1 v$ and $Aw = \lambda_2 w$ with $\lambda_1 < 0 < \lambda_2$. Using Proposition 9 we obtain four semi-lines with the direction of two of these semilines going toward the origin and the direction of the other two semilines going away from the origin. Also we have a dot in the origin representing the only equilibrium point of the system; the point $(0,0)$. The general solution of the system is given by

$$\begin{pmatrix} x(t) \\ y(t) \end{pmatrix} = c_1 e^{\lambda_1 t} v + c_2 e^{\lambda_2 t} w$$

From the expression above we notice that if $c_1 > 0$ and $c_2 > 0$ then the solution lies between the semi-lines l_1 and l_2 because every $\begin{pmatrix} x(t) \\ y(t) \end{pmatrix}$ is a linear combination of the form $s_1 v + s_2 w$ with positive coefficients s_1 and s_2. We notice that when t is big and positive, then the solution is very close to the line $l_2 \cup l_4$ and when t is big and negative, the solution is close to the line $l_1 \cup l_3$.

In general, we have that each orbit that is not a semi-line has an orbit with the shape that looks like a hyperbola (something similar to the graph of the function $y = \frac{1}{x}$) that stays in one of the four quadrants given by the semi-lines l_1, l_2, l_3, l_4 and the orbit has two of these semi-lines as asymptotes. Not only does the orbit approache the asymptote, but the direction of the orbit tries to match the direction of the semi-line that this orbit is converging to. In this case we say that $(0,0)$ is a **saddle**.

Here is an example for this case.

Example 87 *Let us consider the system*

$$\begin{cases} \frac{dx}{dt} &= 4x + y \\ \frac{dy}{dt} &= -5x - 2y \end{cases} \tag{10.2.4}$$

In this case $A = \begin{pmatrix} 4 & 1 \\ -5 & -2 \end{pmatrix}$. *A direct computation shows that the eigenvalues of A are $\lambda_1 = -1$ and $\lambda_2 = 3$. Moreover, if $v = \begin{pmatrix} -1 \\ 1 \end{pmatrix}$ and $w = \begin{pmatrix} -1 \\ 5 \end{pmatrix}$, then $Av = 3v$ and $Aw = -w$. Figure 10.2.4 shows the phase portrait of the differential system.*

10.2.4   Case 4: Two non-real eigenvalues with zero real part

Let us assume that the eigenvalues of A are of the form $\pm \beta i$ with $\beta > 0$. After doing some algebra we have that the general solution of the system is given by

$$\begin{pmatrix} x(t) \\ y(t) \end{pmatrix} = c_1 \begin{pmatrix} d_1 \cos(\beta t) + d_2 \sin(\beta t) \\ d_3 \cos(\beta t) + d_4 \sin(\beta t) \end{pmatrix} + c_2 \begin{pmatrix} e_1 \cos(\beta t) + e_2 \sin(\beta t) \\ e_3 \cos(\beta t) + e_4 \sin(\beta t) \end{pmatrix}$$

The d_i are e_i are found after finding the real and imaginary part of $e^{i\beta t} v$ where v satisfies $Av = \beta i v$. It is evident that the orbits are periodic and they have kind of a circular shape. Sometimes they are perfect circles but most of the time they have an elliptical shape.

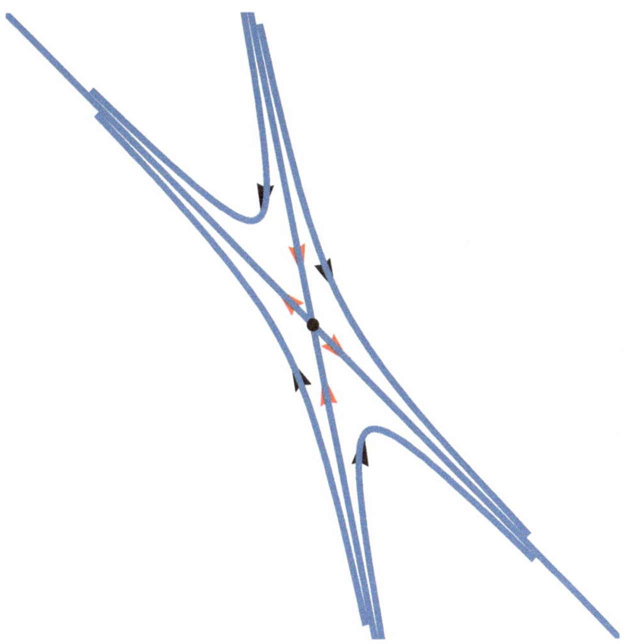

Figure 10.2.4: Phase line example with one positive and one negative eigenvalue. Notice how the semi-lines associated with the eigenvector v have their direction going away from the origin while the semi-lines associated with the eigenvector w have their direction going toward the origin.

A more precise graph can be drawn by noticing that all the orbits are similar to the closed curve that contains the points

$$(b, -a) \rightarrow (0, -\beta) \rightarrow (-b, a) \rightarrow (0, \beta) \rightarrow (b, -a)$$

The direction is given by the order of the points above. Another way to find the direction of the orbits is by selecting the point $(1, 0)$ and evaluating the vector field $V = (ax + by, cx + dy)$ at $(1, 0)$. We notice that $V(1, 0)$ is (a, c). We draw the orbit making sure that the tangent and direction of the orbit at $(1, 0)$ is given by the vector (a, c). For this case we say that $(0, 0)$ is a **center**.

Here is an example for this case.

Example 88 *Let us consider the system*

$$\begin{cases} \frac{dx}{dt} &= 11x + 10y \\ \frac{dy}{dt} &= -13x - 11y \end{cases} \tag{10.2.5}$$

In this case $A = \begin{pmatrix} 11 & 10 \\ -13 & -11 \end{pmatrix}$ and the characteristic equation is given by $\lambda^2 + 9 = 0$ and therefore $\lambda = \pm 3i$. For this example the vector field associated with the differential equation is

$V(x, y) = (11x + 10y, -13x - 11y)$ and therefore $V(1, 0) = (11, -13)$ This indicates that the orbits are going clockwise. Figure 10.2.5 shows the phase portrait of the differential system.

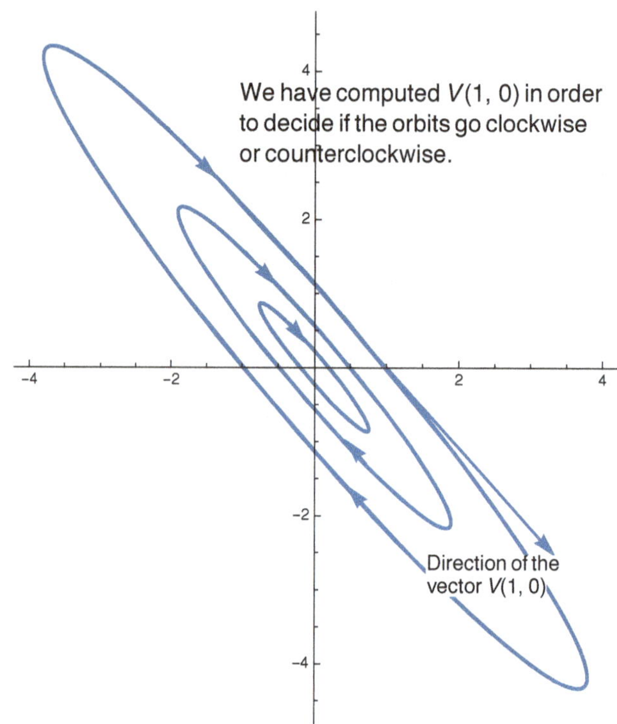

Figure 10.2.5: Example of phase line when $(0, 0)$ is a center.

10.2.5 Case 5: Two non-real eigenvalues with non-zero real part

Let us assume that the eigenvalues of A are of the form $\alpha \pm \beta i$ with $\beta > 0$ and $\alpha \neq 0$. After doing some algebra we have that the general solution of the system is given by

$$\begin{pmatrix} x(t) \\ y(t) \end{pmatrix} = c_1 e^{\alpha t} \begin{pmatrix} d_1 \cos(\beta t) + d_2 \sin(\beta t) \\ d_3 \cos(\beta t) + d_4 \sin(\beta t) \end{pmatrix} + c_2 e^{\alpha t} \begin{pmatrix} e_1 \cos(\beta t) + e_2 \sin(\beta t) \\ e_3 \cos(\beta t) + e_4 \sin(\beta t) \end{pmatrix}$$

The d_i and e_i are found after finding the real and imaginary part of $e^{\alpha t + i\beta t} v$ where v satisfies $Av = (\alpha + i\beta) v$. In this case the orbits are not periodic like in the case when $(0, 0)$ is a center, the case $\alpha = 0$.

This time, when t increases and $\alpha > 0$, the orbits become unbounded while they go around the origin and when $\alpha < 0$ and t increases, the orbits approache $(0,0)$ while they go around the origin. Once again, we can find the direction of the orbit by evaluating the vector field $V = (ax+by, cx+dy)$ at $(1,0)$. We notice that $V(1,0)$ is (a,c). We draw the orbit making sure that the tangent and direction of the orbit at $(1,0)$ is given by the vector (a,c). For this case we say that $(0,0)$ is a **spiral source** if $\alpha > 0$ and we say that $(0,0)$ is a **spiral sink** if $\alpha < 0$.

Here is an example for this case.

Example 89 *Let us consider the system*

$$\begin{cases} \frac{dx}{dt} &= 3x + 5y \\ \frac{dy}{dt} &= -4x - 5y \end{cases} \tag{10.2.6}$$

In this case $A = \begin{pmatrix} 3 & 5 \\ -4 & -5 \end{pmatrix}$ and the characteristic equation is given by $\lambda^2 + 2\lambda + 5 = 0$ and therefore $\lambda = -1 \pm 2i$. For this example the vector field associated with the differential equation is $V(x,y) = (3x + 5y, -4x - 5y)$ and therefore $V(1,0) = (3,-4)$ This indicates that the orbits are going clockwise while they approach $(0,0)$. Figure 10.2.6 shows the phase portrait of the differential system.

10.2.6   Case 6: Repeated non-zero eigenvalues.

In this case we have two cases. One is called proper nodes and the other is called improper node.

Let us assume that the matrix $A = \begin{pmatrix} a & b \\ c & d \end{pmatrix}$ has only one eigenvalue $\lambda \neq 0$. If $b = c = 0$, then all the orbits are semilines with one end at the origin. If $\lambda > 0$, then the direction of each semi-line goes away from the origin and if $\lambda < 0$ the direction of the semi-lines goes toward the origin. In this case the equilibrium $(0,0)$ is called a *proper node*. If either b or c are not zero, then we need to compute the eigenvector v of A and we have the semi-lines $\{tv : t > 0\}$ and $\{tv : t < 0\}$ are orbits.These two semi-lines divide the plane in two half-planes. Any orbit in one of the half planes have the form of a rotated/flipped **J** with the shorter part of the **J** approaching the origin tangent to one of the semi-lines and the longer part extends forever, almost parallel the other semi-line. If we want to decide which semi-line is tangent to the shorter part of the **J** near the origin, we pick a vector V_0 that is not a multiple of the eigenvector v and then we compute $V_1 = (A - \lambda I)V_0$. It turns out that V_1 is a multiple of the eigenvector v. If $\lambda > 0$ then near the origin the orbit that passes through V_0 is almost tangent to the semiline $\{tV_1 : t < 0\}$. If $\lambda < 0$ then, near the origin the orbit that passes through V_0 is almost tangent to the semiline $\{tV_1 : t > 0\}$. In this case the equilibrium point is called an *improper node*.

Here is an example for this case.

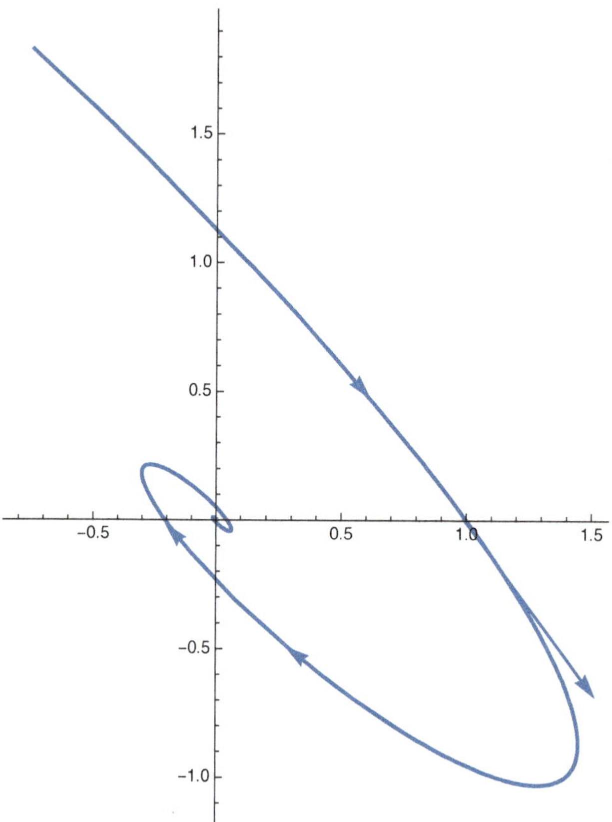

Figure 10.2.6: Phase line when $(0,0)$ is a spiral sink.

Example 90 *Let us consider the system*

$$\begin{cases} \frac{dx}{dt} &= 8x + 3y \\ \frac{dy}{dt} &= -12x - 4y \end{cases} \tag{10.2.7}$$

In this case $A = \begin{pmatrix} 8 & 3 \\ -12 & -4 \end{pmatrix}$ *and the characteristic equation is given by* $\lambda^2 - 4\lambda + 4 = 0$ *and therefore* $\lambda = 2$ *is the only eigenvalue. A direct computation shows that the only eigenvector is* $v = (-1, 2)$. *In order to decide how the orbits approach the origin we take the point* $V_0 = (1, 0)^T$ *(notice that we only need to make sure that* V_0 *is not a multiple of the eigenvector* v*) then,* $V_1 = (A - 2I)V_0 = (6, -12)$. *Since* $\lambda = 2 > 0$, *then the orbit that passes through* $(1, 0)$ *must approach the origin almost tangent to the semiline* $\{t(6, -12) : t < 0\}$. *Figure 10.2.7 shows the phase portrait of the differential system.*

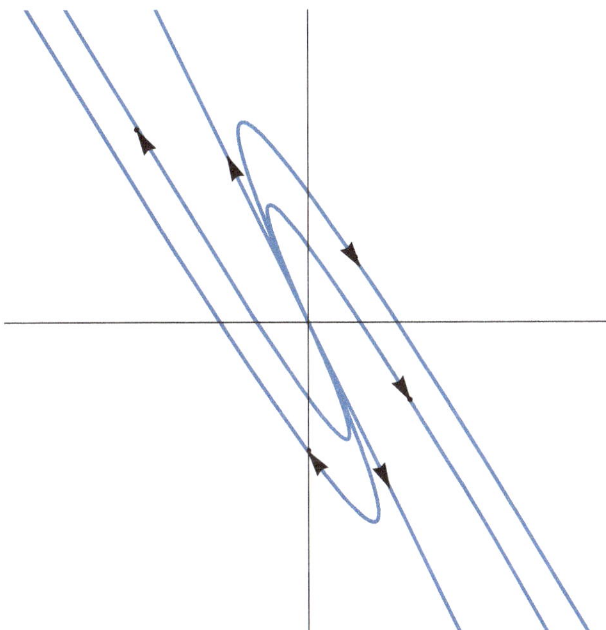

Figure 10.2.7: Phase line when $(0, 0)$ is an improper node.

Another example

Example 91 *Let us consider the system*

$$\begin{cases} \frac{dx}{dt} &= -3x \\ \frac{dy}{dt} &= -3y \end{cases} \tag{10.2.8}$$

In this case $A = \begin{pmatrix} -3 & 0 \\ 0 & -3 \end{pmatrix}$ *and the characteristic equation is given by* $\lambda^2 + 6\lambda + 9 = 0$ *and therefore* $\lambda = -3$ *is the only eigenvalue. Since* $b = c = 0$ *then we have a proper node. Figure 10.2.8 shows the phase portrait of the differential system.*

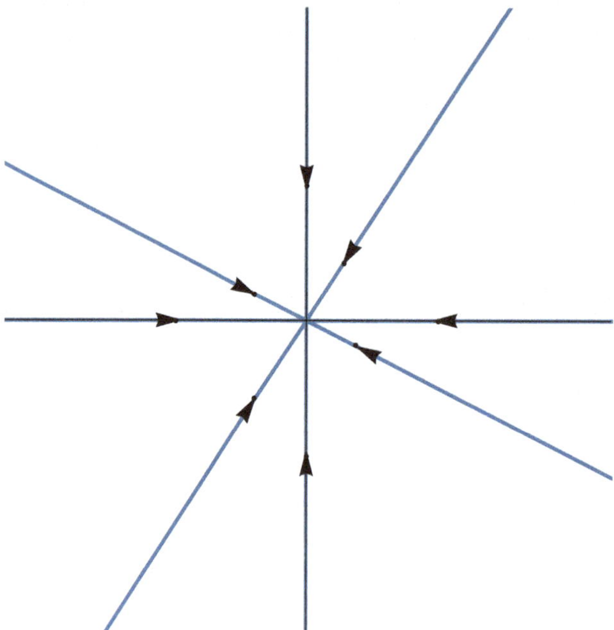

Figure 10.2.8: Phase line when $(0,0)$ is a proper node.

10.3 Homework

For each system sketch the phase portrait making sure to find the eigenvalues of the matrix of the system and the eigenvectors if they are needed. Also, provide the name of the equilibrium point.

1. $\begin{cases} \frac{dx}{dt} &= 6x + 5y \\ \frac{dy}{dt} &= -x \end{cases}$

2. $\begin{cases} \frac{dx}{dt} &= -x - 11y \\ \frac{dy}{dt} &= 3x + 13y \end{cases}$

3. $\begin{cases} \frac{dx}{dt} &= -2x + y \\ \frac{dy}{dt} &= x - 2y \end{cases}$

4. $\begin{cases} \frac{dx}{dt} &= 2x + 3y \\ \frac{dy}{dt} &= -5x - 6y \end{cases}$

5. $\begin{cases} \frac{dx}{dt} &= x + 4y \\ \frac{dy}{dt} &= x + y \end{cases}$

6. $\begin{cases} \frac{dx}{dt} &= 2x + 4y \\ \frac{dy}{dt} &= -3y \end{cases}$

7. $\begin{cases} \frac{dx}{dt} &= x + 8y \\ \frac{dy}{dt} &= -2x - 7y \end{cases}$

8. $\begin{cases} \frac{dx}{dt} &= 3x + 8y \\ \frac{dy}{dt} &= 3y \end{cases}$

9. $\begin{cases} \frac{dx}{dt} &= -4x \\ \frac{dy}{dt} &= -4y \end{cases}$

10. $\begin{cases} \frac{dx}{dt} &= 7x + 2y \\ \frac{dy}{dt} &= -5x + y \end{cases}$

11. $\begin{cases} \frac{dx}{dt} &= 3x + 2y \\ \frac{dy}{dt} &= -5x - 3y \end{cases}$

12. $\begin{cases} \frac{dx}{dt} &= x + y \\ \frac{dy}{dt} &= -x + y \end{cases}$

10.4 Answers

1. Eigenvalues with eigenvectors: 5 with eigenvector $(-5, 1)^T$ and 1 with eigenvector $(-1, 1)^T$. In this example $(0, 0)$ is a source.

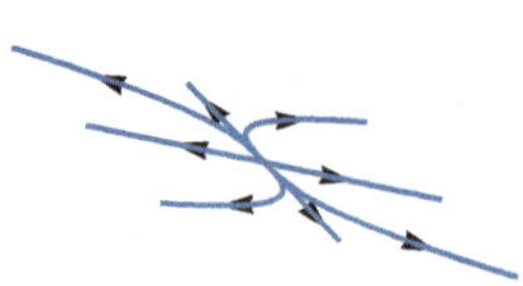

2. Eigenvalues with eigenvectors: 10 with eigenvector $(-1, 1)^T$ and 2 with eigenvector $(-11, 3)^T$. In this example $(0, 0)$ is a source.

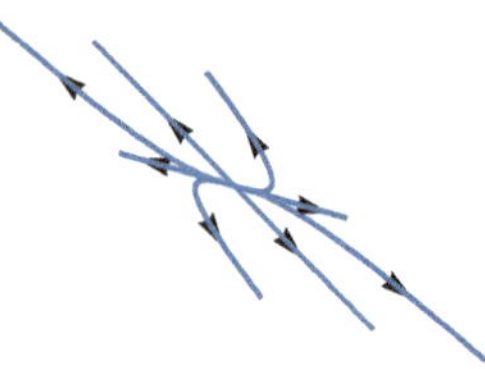

3. Eigenvalues with eigenvectors: -3 with eigenvector $(-1, 1)^T$ and -1 with eigenvector $(1, 1)^T$. In this example $(0, 0)$ is a sink.

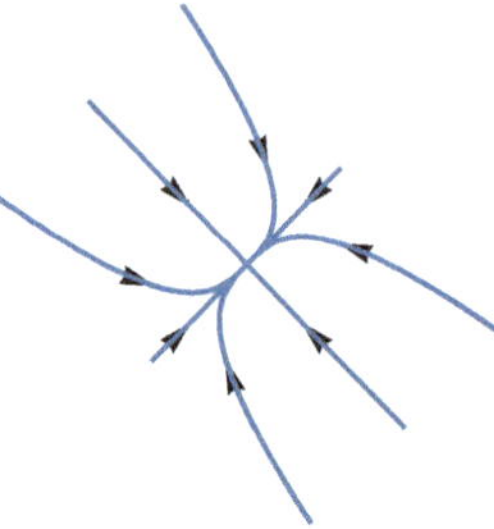

4. Eigenvalues with eigenvectors: -3 with eigenvector $(-3, 5)^T$ and -1 with eigenvector $(-1, 1)^T$. In this example $(0, 0)$ is a sink.

5. Eigenvalues with eigenvectors: 3 with eigenvector $(2, 1)^T$ and -1 with eigenvector $(-2, 1)^T$. In this example $(0, 0)$ is a saddle.

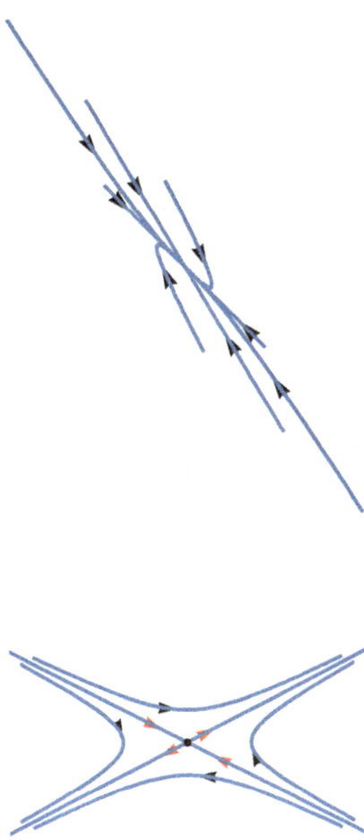

6. Eigenvalues with eigenvectors: -3 with eigenvector $(-4,5)^T$ and 2 with eigenvector $(1,0)^T$. In this example $(0,0)$ is a saddle.

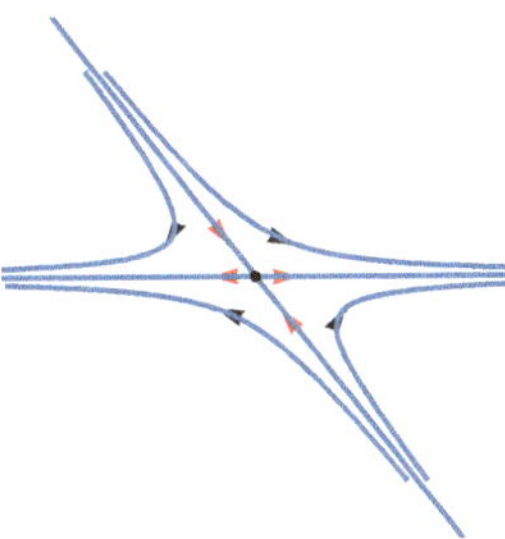

7. Eigenvalues with eigenvectors: -3 with eigenvector $(-2,1)^T$, there is only one eigenvalue. In this example $(0,0)$ is an improper node.

8. Eigenvalues with eigenvectors: 3 with eigenvector $(1,0)^T$, there is only one eigenvalue. In this example $(0,0)$ is an improper node.

9. Eigenvalues -4, diagonal matrix . In this example $(0,0)$ is a proper node.

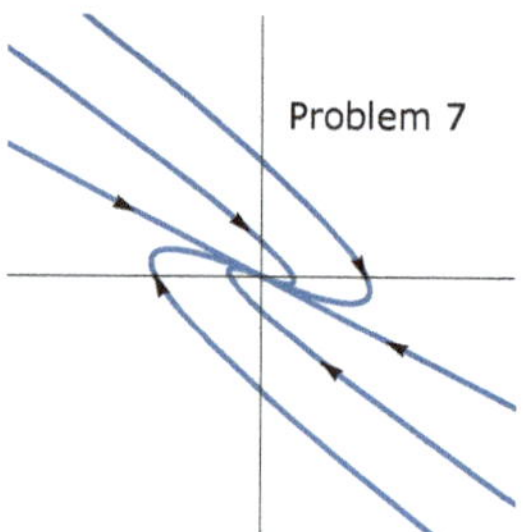
Problem 7

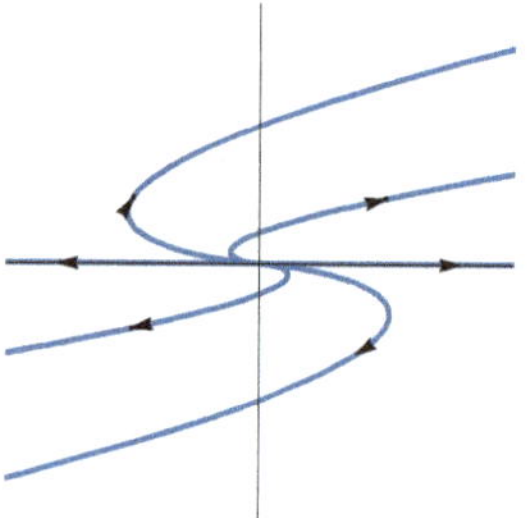

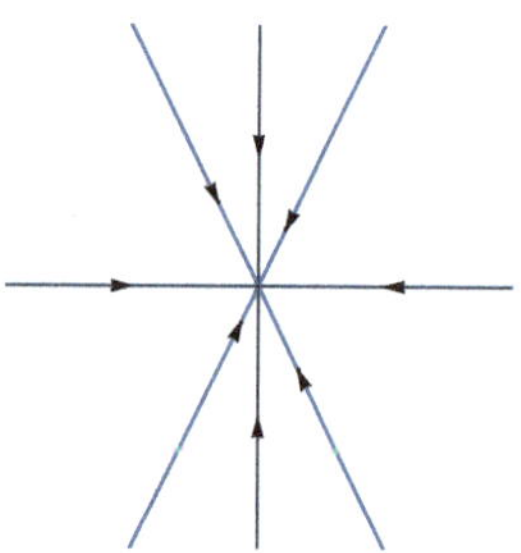

10. Eigenvalues $4 \pm i$, the spirals go clockwise. In this example $(0,0)$ is a spiral source

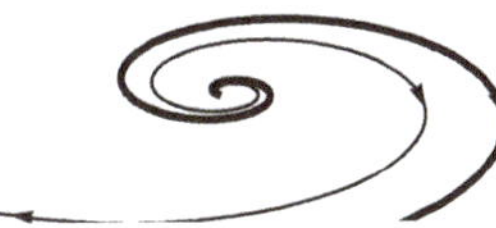

11. Eigenvalues $\pm i$, the orbits go clockwise. In this example $(0,0)$ is a center

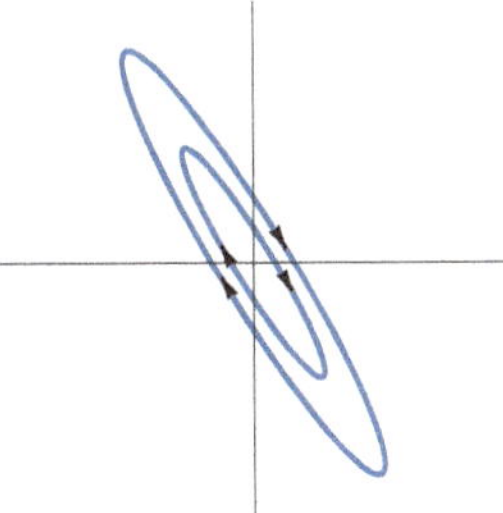

12. Eigenvalues $1 \pm i$, the spirals go clockwise. In this example $(0,0)$ is a spiral source

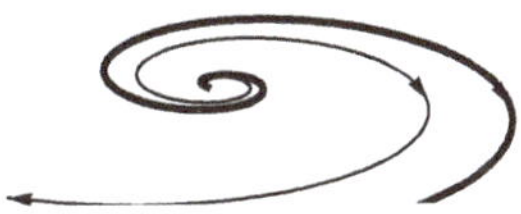

chapter 11

Laplace transform part 1

11.1 Definition and some properties of the Laplace transform

The Laplace transform is an operator that acts on functions. In this sense it is similar to the derivative operator. Notice that the derivative operator acts on the function $f(t) = t^2$ and it changes it into the function $f'(t) = 2t$.

> One of the differences between the derivative operator and the Laplace transform is that the Laplace transform operator takes a function that depends on t and it transforms it into a function that depends on s. For example, we will see that the Laplace transform of $f(t) = t^2$ is the function $F(s) = \frac{2}{s^3}$.

This is the definition of Laplace transform,

> **Definition 15** *For any function $f(t)$ defined on the interval $(0, \infty)$ we define the Laplace transform of $f(t)$, denoted by $F(s) = \mathcal{L}(f(t))$ as the function*
>
> $$\mathcal{L}(f(t))(s) = F(s) = \int_0^\infty f(t)e^{-st}dt$$

We point out that the domain of the function $\mathcal{L}(f(t))$ depends on the function $f(t)$, this domain is usually an interval of the form $[a, \infty)$, this is, the set $\{s : s \geq a\}$. Also, for some functions $f(t)$ the Laplace Transform may not exist due to the fact that the integral could diverge for all values of s.

> We will not be concerned about the domain of $\mathcal{L}(f(t))$ and, in the examples that we will be dealing with, we will also not be concerned about the infinity in the improper integral, due to the fact that in all our examples the antiderivative evaluated at infinity is zero (or to be more precise, the limit of the antiderivative when t goes to infinity vanishes).

Example 92 *If $f(t) = 3$, then*

$$\mathcal{L}(f(t)) = \int_0^\infty 3e^{-st}dt = \frac{3}{-s}e^{-st}\Big|_{t=0}^{t=\infty} = 0 - \frac{3}{-s} = \frac{3}{s}$$

Notice that we are skipping some steps in the previous computation. To be mathematically precise we would have needed to write

$$\begin{aligned}
\mathcal{L}(3) &= \int_0^\infty 3e^{-st}dt = \lim_{b\to\infty}\int_0^b 3e^{-st}dt \\[2mm]
&= \lim_{b\to\infty}\frac{3}{-s}e^{-st}\Big|_{t=0}^{t=b} = \lim_{b\to\infty}\left(\frac{3}{-s}e^{-sb} - \frac{3}{-s}e^{-s0}\right) \\[2mm]
&= 0 - \frac{3}{-s} \quad, \text{ assuming } s > 0 \\[2mm]
&= \frac{3}{s}
\end{aligned}$$

Referring to the comment before the example, in this case the antiderivative is the function $\frac{3}{-s}e^{-st}$ and, as we mentioned earlier, the limit of this antiderivative when t goes to infinity vanishes (assuming $s > 0$).

In the same way we prove that $\mathcal{L}(3) = \frac{3}{s}$ we can show that

Proposition 10

$$\mathcal{L}(a) = \frac{a}{s} \quad \text{ for any constant } a$$

Example 93 *If $f(t) = e^{2t}$, then*

$$\mathcal{L}(f(t)) = \int_0^\infty e^{2t}e^{-st}dt = \int_0^\infty e^{2t-st}dt = \frac{1}{2-s}e^{2t-st}\Big|_{t=0}^{t=\infty} = 0 - \frac{1}{2-s} = \frac{1}{s-2}$$

In the same way we prove that $\mathcal{L}(e^{2t}) = \frac{1}{s-2}$ we can show that,

Proposition 11

$$\mathcal{L}(e^{at}) = \frac{1}{s-a} \quad \text{ for any constant } a$$

Example 94 *If $f(t) = t$, then $\mathcal{L}(f(t)) = \int_0^\infty te^{-st}dt$. We will do integration by parts to solve the integral. Making $u = t$ and $dv = e^{-st}dt$ we get that $du = dt$ and $v = \frac{e^{-st}}{-s}$, then,*

$$\int te^{-st}dt = -\frac{t}{s}e^{-st} - \int \frac{1}{-s}e^{-st}\,dt = -t\frac{1}{s}e^{-st} - \frac{1}{s^2}e^{-st}$$

Therefore,

$$\int_0^\infty te^{-st}dt = \left(-\frac{t}{s}e^{-st} - \frac{1}{s^2}e^{-st}\right)\Big|_{t=0}^{t=\infty} = 0 - \left(-\frac{1}{s^2}\right) = \frac{1}{s^2}$$

In the previous computation we have used the fact that for any $s > 0$,

$$\lim_{t \to \infty} t \frac{1}{s} e^{-st} = \lim_{t \to \infty} \frac{t}{s e^{st}} = 0$$

which can be shown using L'Hospital's rule.

In the same way the derivative of a sum of two functions is the sum of their derivatives, we have that the Laplace transform of the sum of two functions is the sum of their Laplace transforms. Also, the Laplace transform of the product of a function with a constant, is the product of the constant with the Laplace transform of the function.

Proposition 12

$$\mathcal{L}(f(t) + g(t)) = \mathcal{L}(f(t)) + \mathcal{L}(g(t)) \quad and \quad \mathcal{L}(cf(t)) = c\mathcal{L}(f(t))$$

Example 95

$$\mathcal{L}(2 + 3e^{-6t} + 4t) = \frac{2}{s} + 3\mathcal{L}(e^{-6t}) + 4\mathcal{L}(t) = \frac{2}{s} + \frac{3}{s+6} + \frac{4}{s^2}$$

The next theorem is one of the most important properties for the Laplace transform.

Theorem 12

$$\mathcal{L}(\frac{dy}{dt}) = s\mathcal{L}(y) - y(0)$$

Proof We have that $\mathcal{L}(\frac{dy}{dt}) = \int_0^\infty y'(t)e^{-st}\,dt$. Using integration by parts with $u = e^{-st}$ and $dv = y'(t)dt$, then $v(t) = y(t)$ and $du = -se^{-st}$, we obtain that

$$\int_0^\infty y'(t)e^{-st}\,dt = e^{-st}y(t)\Big|_{t=0}^{t=\infty} + \int_0^\infty y(t)se^{-st}\,dt$$

As mentioned before we will be assuming that the limit when t goes to infinity of $e^{-st}y(t)$ is zero. Then, we have that

$$\int_0^\infty y'(t)e^{-st}\,dt = 0 - (e^{-s\times0}y(0)) + s\int_0^\infty y(t)e^{-st}\,dt = -y(0) + s\mathcal{L}(y)$$

As an application of the previous theorem we can compute the Laplace transform of the function t^n for any positive integer n. We have

Example 96 *Since the derivative of $f(t) = t^2$ is $f'(t) = 2t$, then using the formula $\mathcal{L}(\frac{dy}{dt}) = s\mathcal{L}(y) - y(0)$ we obtain that*

$$2\frac{1}{s^2} = \mathcal{L}(2t) = 0 + s\mathcal{L}(t^2)$$

We obtain that $\mathcal{L}(t^2) = \frac{2}{s^3}$. Notice that we have used that $\mathcal{L}(t) = \frac{1}{s^2}$. We can do the same argument with the function $f(t) = t^3$. We have that $f'(t) = 3t^2$ and therefore

$$3\frac{2}{s^3} = \mathcal{L}(3t^2) = 0 + s\mathcal{L}(t^3)$$

and therefore $\mathcal{L}(t^3) = \frac{2\times 3}{s^4}$. Continuing with this process we can show that

$$\mathcal{L}(t^n) = \frac{n!}{s^{n+1}}$$

11.2   Laplace transform inverse

Definition 16 *We say that $f(t)$ is the Laplace transform inverse of $F(s)$ if $\mathcal{L}(f(t)) = F(s)$. When this happens we write*

$$f(t) = \mathcal{L}^{-1}(F(s))$$

These are some examples,

Example 97 *Since $\mathcal{L}(a) = \frac{a}{s}$ then,*

$$\mathcal{L}^{-1}(\frac{a}{s}) = a$$

In the same way, we have that

$$\mathcal{L}^{-1}(\frac{1}{s^{n+1}}) = \frac{1}{n!}\,t^n \quad and \quad \mathcal{L}^{-1}(\frac{1}{s-a}) = e^{at}$$

One of the major tools used to find Laplace transform and Laplace transform inverse is the ability to rewrite an expression.

Example 98

$$\mathcal{L}^{-1}(\frac{3}{4+5s}) = \mathcal{L}(\frac{3}{5(\frac{4}{5}+s)}) = \mathcal{L}(\frac{\frac{3}{5}}{s-(-\frac{4}{5})}) = \frac{3}{5}e^{-\frac{4}{5}t}$$

Here is another typical example where the key part is to rewrite the function,

Example 99 *In order to compute the Laplace transform inverse of the function $F(s) = \frac{4+6s}{s^2-4}$ we do partial fractions due to the fact that we know how to compute the Laplace transform inverse of expressions of the form $\frac{a}{bs+c}$. Using partial fractions we have that*

$$\frac{4+6s}{s^2-4} = \frac{4+6s}{(s-2)(s+2)} = \frac{A}{s-2} + \frac{B}{s+2} = \frac{A(s+2)+B(s-2)}{(s-2)(s+2)}$$

Since the two denominators of the rational functions above are the same then

$$4+6s = A(s+2) + B(s-2)$$

We get two equations by noticing that the constant coefficient of the LHS is 4, while the constant coefficient of the RHS is $2A - 2B$. Also the coefficient of s on the LHS is 6 while the coefficient of s

on the RHS is $A + B$. Solving the system $\begin{cases} 2A - 2B & = 4 \\ A + B & = 6 \end{cases}$ *we obtain that $A = 4$ and $B = 2$, this is*

$$\frac{4 + 6s}{s^2 - 4} = \frac{4}{s - 2} + \frac{2}{s + 2} = \mathcal{L}(4e^{2t} + 2e^{-2t})$$

This is, $\boxed{\mathcal{L}^{-1}\left(\frac{4 + 6s}{s^2 - 4}\right) = 4e^{2t} + 2e^{-2t}}$

Example 100 *In order to compute the Laplace transform inverse of the function $F(s) = \frac{2s^3 + s^2 + 3s + 2}{s^3(s^2 - 3s + 2)}$ we do partial fractions due to the fact that we know how to compute the Laplace transform inverse of expressions of the form $\frac{a}{bs + c}$ and $\frac{1}{s^n}$. Using partial fractions we have that $\frac{2s^3 + s^2 + 3s + 2}{s^3(s^2 - 3s + 2)}$ is equal to*

$$\frac{2s^3 + s^2 + 3s + 2}{s^3(s - 1)(s - 2)} = \frac{A}{s} + \frac{B}{s^2} + \frac{C}{s^3} + \frac{D}{s - 1} + \frac{E}{s - 2} =$$
$$\frac{As^2(s - 1)(s - 2) + Bs(s - 1)(s - 2) + C(s - 1)(s - 2) + Ds^3(s - 2) + Es^3(s - 1)}{s^3(s - 1)(s - 2)}$$

Since the two denominators of the rational functions above are the same then

$$As^2(s-1)(s-2) + Bs(s-1)(s-2) + C(s-1)(s-2) + Ds^3(s-2) + Es^3(s-1) = 2s^3 + s^2 + 3s + 2$$

We get the following system of equations by comparing the coefficients of s^4, s^3, s^3, s^2, s and the constant coefficients on both sides of the equation above

$$\begin{cases} A + D + E & = 0 \\ -3A + B - 2D - E & = 2 \\ 2A - 3B + C & = 1 \\ 2B - 3C & = 3 \\ 2C = 2 \end{cases}$$

Since the solution of the system above is $A = \frac{9}{2}, B = 3, C = 1, D = -8, E = \frac{7}{2}$, then

$$\frac{2s^3 + s^2 + 3s + 2}{s^3(s^2 - 3s + 2)} = \frac{\frac{9}{2}}{s} + \frac{3}{s^2} + \frac{1}{s^3} - \frac{8}{s - 1} + \frac{\frac{7}{2}}{s - 2}$$
$$= \mathcal{L}\left(\frac{9}{2} + 3t + \frac{t^2}{2} - 8e^t + \frac{7}{2}e^{2t}\right)$$

This is, $\boxed{\mathcal{L}^{-1}\left(\frac{2s^3 + s^2 + 3s + 2}{s^3(s^2 - 3s + 2)}\right) = \frac{9}{2} + 3t + \frac{t^2}{2} - 8e^t + \frac{7}{2}e^{2t}}$

Notice that partial fractions is not going to help us to compute $\mathcal{L}^{-1}\left(\frac{1}{(s-2)^2}\right)$. The following theorem will take care of this Laplace transform inverse.

Theorem 13 *If $F(s) = \mathcal{L}(f(t))$, then $\mathcal{L}(tf(t)) = -F'(s)$.*

Proof The equation $F(s) = \mathcal{L}(f(t))$ can be written as $F(s) = \int_0^\infty f(t)e^{-st}\,dt$. Taking the derivative with respect to s on both sides of this equation gives us that

$$F'(s) = -\int_0^\infty tf(t)e^{-st}\,dt = -\mathcal{L}(tf(t))$$

Example 101 *Since* $(\frac{1}{s-3})' = -\frac{1}{(s-3)^2}$, *then* $\mathcal{L}(te^{3t}) = -(\frac{1}{s-3})' = \frac{1}{(s-3)^2}$. *In general we have that*

$$\mathcal{L}(te^{at}) = \frac{1}{(s-a)^2}$$

11.3   Solving differential equations using Laplace transforms

This section shows how to use the Laplace transform to solve differential equations. Let us consider the initial value problem

$$\frac{dy}{dt} + 4y = 6e^{2t}, \quad y(0) = 3$$

Using the formula $\mathcal{L}(\frac{dy}{dt}) = s\mathcal{L}(y) - y(0)$, we obtain, by applying the Laplace transform on both sides of the differential equation, that

$$s\mathcal{L}(y) - 3 + 4\mathcal{L}(y) = \frac{6}{s-2}$$

From the equation above we obtain that $(s+4)\mathcal{L}(y) = 3 + \frac{6}{s-2} = \frac{3s-6+6}{s-2} = \frac{3s}{s-2}$ and therefore,

$$\mathcal{L}(y) = \frac{3s}{(s-2)(s+4)}$$

This is the first step needed to solve a differential equation using Laplace transform. We apply Laplace transform on both sides of the equation and then we solve for $\mathcal{L}(y)$. The next step is to compute the Laplace transform inverse of the expression in front of $\mathcal{L}(y)$. In this case, in order to compute the Laplace transform inverse of $\frac{3s}{(s-2)(s+4)}$ we need to use partial fractions. We have that

$$\frac{3s}{(s-2)(s+4)} = \frac{A}{s-2} + \frac{B}{s+4} = \frac{A(s+4) + B(s-2)}{(s-2)(s+4)}$$

The system of equation in this case is $\begin{cases} A + B &= 3 \\ 4A - 2B &= 0 \end{cases}$. Since the solution is $A = 1$ and $B = 2$, we get that

$$\frac{3s}{(s-2)(s+4)} = \frac{1}{s-2} + \frac{2}{s+4} = \mathcal{L}(e^{2t} + 2e^{-4t})$$

Moreover we have that

$$\mathcal{L}(y) = \frac{3s}{(s-2)(s+4)} = \frac{1}{s-2} + \frac{2}{s+4} = \mathcal{L}(e^{2t} + 2e^{-4t})$$

and therefore, $\boxed{y = e^{2t} + 2e^{-4t}}$

> **Remark 7** *Notice that the steps to solve a differential equation using Laplace transform are:*
>
> - *Compute the Laplace transform on both sides of the equation.*
>
> - *Solve for $\mathcal{L}(y)$. You need to get an equation of the form $\mathcal{L}(y) = F(s)$.*
>
> - *Compute $\mathcal{L}^{-1}(F(s))$. We have that $y = \mathcal{L}^{-1}(F(s))$.*

Example 102 *Let us consider the following initial value problem*

$$\frac{dy}{dt} + 3y = 2 + 3t, \quad y(0) = -2$$

Applying the Laplace transform on both sides of the equation we get

$$s\mathcal{L}(y) + 2 + 3\mathcal{L}(y) = \frac{2}{s} + \frac{3}{s^2}$$

then

$$(s+3)\mathcal{L}(y) = -2 + \frac{2}{s} + \frac{3}{s^2} = \frac{-2s^2 + 2s + 3}{s^2}$$

and then

$$\mathcal{L}(y) = \frac{-2s^2 + 2s + 3}{s^2(s+3)} = \frac{A}{s} + \frac{B}{s^2} + \frac{C}{s+3} = \frac{As(s+3) + B(s+3) + Cs^2}{s^2(s+3)}$$

We can find the values of A, B and C by solving the following system

$$\begin{cases} A + C & = -2 \\ 3A + B & = 2 \\ 3B & = 3 \end{cases}$$

The solution is $B = 1$, $A = \frac{1}{3}$ and $C = -\frac{7}{3}$. Then,

$$\mathcal{L}(y) = \frac{-2s^2 + 2s + 3}{s^2(s+3)} = \frac{\frac{1}{3}}{s} + \frac{1}{s^2} - \frac{\frac{7}{3}}{s+3} = \mathcal{L}(\frac{1}{3} + t - \frac{7}{3}e^{-3t})$$

and therefore, $\boxed{y = \frac{1}{3} + t - \frac{7}{3}e^{-3t}}$

Example 103 *Let us consider the following initial value problem*

$$\frac{dy}{dt} - 4y = 2e^{4t}, \quad y(0) = 2$$

Applying the Laplace transform to both sides of the equation we get

$$s\mathcal{L}(y) - 2 - 4\mathcal{L}(y) = \frac{2}{s-4}$$

then

$$(s - 4)\mathcal{L}(y) = 2 + \frac{2}{s - 4}$$

and then

$$\mathcal{L}(y) = \frac{2}{s - 4} + \frac{2}{(s - 4)^2} = \mathcal{L}(2e^{4t} + 2te^{4t})$$

and therefore, $\boxed{y = 2e^{4t} + 2te^{4t}}$

The following theorem will be used to solve second order differential equations and for now it will be useful to compute the Laplace of the functions $y = \sin(\omega t)$ and $y = \cos(\omega t)$.

Theorem 14

$$\mathcal{L}(\frac{d^2 y}{dt^2}) = s^2 \mathcal{L}(y) - y(0)s - y'(0)$$

Proof Using the formula $\mathcal{L}(\frac{dz}{dt}) = s\mathcal{L}(z) - z(0)$ with $z(t) = y'(t)$ we obtain that

$$\mathcal{L}(\frac{d^2 y}{dt^2}) = s\mathcal{L}(\frac{dy}{dt}) - y'(0) = s\left(s\mathcal{L}(y) - y(0)\right) - y'(0) = s^2\mathcal{L}(y) - y(0)s - y'(0)$$

Example 104 *The function* $y = \sin(7t)$ *satisfies the initial value problem*

$$\frac{d^2 y}{dt^2} + 49y = 0 \quad y(0) = 0 \quad y'(0) = 7$$

Applying the formula for the Laplace transform on both sides of the equation above we obtain that

$$s^2 \mathcal{L}(y) - 7 + 49\mathcal{L}(y) = 0$$

and therefore, $(s^2 + 49)\mathcal{L}(y) = 7$. *This is,*

$$\mathcal{L}(y) = \frac{7}{s^2 + 49}$$

In the same way we can prove that for any $\omega > 0$,

$$\mathcal{L}(\sin(\omega t)) = \frac{\omega}{s^2 + \omega^2}$$

We have that $y = \cos(\omega t)$ *satisfies the initial value problem*

$$\frac{d^2 y}{dt^2} + \omega^2 y = 0 \quad y(0) = 1 \quad y'(0) = 0$$

Applying the formula for the Laplace transform on both sides of equation above we obtain that

$$s^2 \mathcal{L}(y) - s + \omega^2 \mathcal{L}(y) = 0$$

and therefore,

$$\mathcal{L}(\cos(\omega t)) = \frac{s}{s^2 + \omega^2}$$

Example 105 *To compute the Laplace transform inverse of the function* $F(s) = \frac{3}{s^2+16}$ *we rewrite*

$F(s)$ *as* $\frac{3}{4}\frac{4}{s^2+16}$ *and therefore* $F(s) = \mathcal{L}(\frac{3}{4}\sin(4t))$, *this is* $\boxed{\mathcal{L}^{-1}\left(\dfrac{3}{s^2+16}\right) = \dfrac{3}{4}\sin(4t)}$.

Example 106 *To compute the Laplace transform inverse of the function* $F(s) = \frac{3s+2}{s^2+7}$ *we rewrite* $F(s)$ *as*

$$F(s) = \frac{3s}{s^2+7} + \frac{2}{s^2+7} = 3\frac{s}{s^2+7} + \frac{2}{\sqrt{7}}\frac{\sqrt{7}}{s^2+7} = \mathcal{L}\left(3\cos(\sqrt{7}t) + \frac{2}{\sqrt{7}}\sin(\sqrt{7}t)\right)$$

and therefore $\boxed{\mathcal{L}^{-1}\left(\dfrac{3s+2}{s^2+7}\right) = 3\cos(\sqrt{7}t) + \dfrac{2}{\sqrt{7}}\sin(\sqrt{7}t)}$.

Now let us solve a differential equations that has a sine function.

Example 107 *transform on Let us consider the following initial value problem*

$$\frac{dy}{dt} + y = 5\sin(2t), \quad y(0) = 3$$

Applying the Laplace transform on both sides of the equation we get

$$s\mathcal{L}(y) - 3 + \mathcal{L}(y) = 5\frac{2}{s^2+4}$$

then

$$(s+1)\mathcal{L}(y) = 3 + \frac{10}{s^2+4} = \frac{3s^2+22}{s^2+4}$$

and then

$$\mathcal{L}(y) = \frac{3s^2+22}{(s^2+4)(s+1)} = \frac{A}{s+1} + \frac{Bs+C}{s^2+4} = \frac{A(s^2+4) + (Bs+C)(s+1)}{(s^2+4)(s+1)}$$

We can find the values of A, B *and* C *by solving the following system*

$$\begin{cases} A + B & = 3 \\ B + C & = 0 \\ 4A + C & = 22 \end{cases}$$

The solution is $A = 5$, $B = -2$ *and* $C = 2$. *Then,*

$$\mathcal{L}(y) = \frac{5}{s+1} + \frac{-2s+2}{s^2+4} = \frac{5}{s+1} - 2\frac{s}{s^2+4} + \frac{2}{s^2+4} = \mathcal{L}(5e^{-t} - 2\cos(2t) + \sin(2t))$$

and therefore, $\boxed{y - 5e^{-t} - 2\cos(2t) + \sin(2t)}$

11.4 Partial fraction decomposition

Partial fraction decomposition provides a way to rewrite a rational function (a quotient of two polynomials) as a sum of "simpler" rational functions.

> **Definition 17** *A polynomial $p(s) = s^2 + bs + c$ of degree two is irreducible if $b^2 - 4c < 0$, in other words, $p(s)$ is irreducible if the equation $s^2 + bs + c = 0$ has no real solutions.*

In the same way any positive integer can be written as a product of prime numbers, any polynomial can be written as a product of linear polynomials and irreducible polynomials of degree 2. Some examples of both facts are: $12 = 3 \times 2 \times 2 = 3 \times 2^2$, $s^2 - 9$ which is not irreducible can be written as $(s-3)(s+3)$ or $s^3 - s^2 - s - 15 = (s-3)(s^2+2s+5)$, notice that s^2+2s+5 is irreducible because $2^2 - 4 \times 5 < 0$. At this point we would like to point out that factoring a polynomial may be very challenging and is the more difficult part of the process of partial fraction decomposition. For the examples that we will be working on in these notes, the factoring is either given or is easy to do. Let us write the precise statement of the factorization that we have just mentioned.

> **Theorem 15** *Any polynomial with real coefficients $q(s) = s^n + a_{n-1}s^{n-1} + \cdots + a_1 s + a_0$ can be written in a unique way as*
>
> $$q(s) = (s - r_1)^{m_1} \ldots (s - r_k)^{m_k} (s^2 + b_1 s + c_1)^{l_1} \ldots (s^2 + b_u s + c_u)^{l_u}$$
>
> *where the r_i are different real numbers and the (b_i, c_i) are different pairs of real numbers that satisfy $b_i^2 - 4c_i < 0$; this is, the polynomials $s^2 - b_i s - c_i$ are irreducible.*

Example 108 *The polynomial $q(s) = s^3 - 4s^2$ can be written as $q(s) = s^2(s-1)$. Then, comparing with the equation in Theorem (15) we can take $r_1 = 0$, $m_1 = 2$, $r_2 = 1$ and $m_2 = 1$. Also in this case $k = 2$ and $u = 0$*

Example 109 *Comparing with Equation (15), for the polynomial $q(s) = s^3(s + 2)(s^2 + 2s + 9)^2(s^2 + 3)$ we can take $r_1 = 0$, $m_1 = 3$, $r_2 = -2$, $m_2 = 1$, $b_1 = 2$, $c_1 = 9$, $l_1 = 2$, $b_2 = 0$, $c_2 = 3$ and $l_2 = 1$. In this case $k = 2$ and $u = 2$.*

Theorem 16 *If* $q(s) = (s - r_1)^{m_1} \ldots (s - r_k)^{m_k}(s^2 + b_1 s + c_1)^{l_1} \ldots (s^2 + b_u s + c_u)^{l_u}$ *with the* r_i *different real numbers and the* (b_i, c_i) *different pairs of real numbers satisfying* $b_i^2 - 4c_i < 0$ *and* $q(s)$ *is a polynomial with degree smaller than the degree of* $p(s)$*, then*

$$
\begin{aligned}
\frac{p(s)}{q(s)} \;=\;\; & \frac{A_{11}}{s - r_1} + \cdots + \frac{A_{1m_1}}{(s - r_1)^{m_1}} + \frac{A_{21}}{s - r_2} + \cdots + \frac{A_{2m_2}}{(s - r_2)^{m_2}} + \\[2mm]
& \cdots + \frac{A_{k1}}{s - r_k} + \cdots + \frac{A_{km_k}}{(s - r_k)^{m_k}} + \frac{B_{11}s + C_{11}}{s^2 + b_1 s + c_1} + \cdots + \\[2mm]
& \frac{B_{1l_1}s + C_{1l_1}}{(s^2 + b_1 s + c_1)^{l_1}} + \frac{B_{21}s + C_{21}}{s^2 + b_2 s + c_2} + \cdots + \frac{B_{2l_2}s + C_{2l_2}}{(s^2 + b_2 s + c_2)^{l_2}} \\[2mm]
& + \cdots + \frac{B_{u1}s + C_{u1}}{s^2 + b_u s + c_u} + \cdots + \frac{B_{ul_u}s + C_{ul_u}}{(s^2 + b_u s + c_u)^{l_u}}
\end{aligned}
$$

for some constants A_{ij}*,* B_{ij} *and* C_{ij}*.*

Remark 8 *The theorem establishes the existence of the constants* A_{ij}*,* B_{ij} *and* C_{ij} *that make the rewriting of the expression* $\frac{p(s)}{q(s)}$ *true. When we want to find the constants, the problem reduces to that of solving a linear system of equations. An important check point is that the number of unknowns in this system, this is the number of* A_{ij}*,* B_{ij} *and* C_{ij} *equals the degree of the polynomial* $q(s)$

11.5 Table of Laplace transforms

1.
$$\mathcal{L}(1) = \frac{1}{s}$$

2.
$$\mathcal{L}(e^{at}) = \frac{1}{s-a}$$

3.
$$\mathcal{L}(t) = \frac{1}{s^2}$$

4.
$$\mathcal{L}(t^n) = \frac{n!}{s^{n+1}}$$

5.
$$\mathcal{L}(u_a(t)) = \frac{e^{-as}}{s}$$

6.
$$\mathcal{L}(\sin(\omega t)) = \frac{\omega}{s^2 + \omega^2}$$

7.
$$\mathcal{L}(\cos(\omega t)) = \frac{s}{s^2 + \omega^2}$$

8.
$$\mathcal{L}(e^{at}\sin(\omega t)) = \frac{\omega}{(s-a)^2 + \omega^2}$$

9.
$$\mathcal{L}(e^{at}\cos(\omega t)) = \frac{s-a}{(s-a)^2 + \omega^2}$$

10.
$$\mathcal{L}(u_a(t)f(t-a)) = e^{-as}\mathcal{L}(f(t))$$

11.
$$\mathcal{L}(\delta_a(t)) = e^{-as}$$

12.
$$\mathcal{L}(\frac{dy}{dt}) = s\mathcal{L}(y) - y(0)$$

13.
$$\mathcal{L}(\frac{d^2y}{dt^2}) = s^2\mathcal{L}(y) - y(0)s - y'(0)$$

14. .
$$\text{If } \mathcal{L}(f) = F(s), \text{ then } \mathcal{L}(tf(t)) = -\frac{dF}{ds}, \text{ in particular } \mathcal{L}(te^{at}) = \frac{1}{(s-a)^2}$$

11.6   Homework

1. Compute the Laplace transform of $f(t) = 6$ using the definition. Answer:

$$\mathcal{L}(6) = \int_0^\infty 6e^{-st}dt = \frac{6}{-s}e^{-st}\Big|_{t=0}^{t=\infty} = 0 - \frac{6}{-s} = \frac{6}{s}$$

2. Compute the Laplace transform of $f(t) = 3e^{4t}$ using the definition. Answer: $\mathcal{L}(3e^{4t}) =$

$$\int_0^\infty 3e^{4t}e^{-st}dt = \int_0^\infty 3e^{(4-s)t}dt = \frac{3}{4-s}e^{(4-s)t}\Big|_{t=0}^{t=\infty} = 0 - \frac{3}{4-s} = \frac{3}{s-4}$$

3. Solve using Laplace transform, $\frac{dy}{dt} = 3y + 5$, $y(0) = 2$. Answer: $y = \frac{11e^{3t}}{3} - \frac{5}{3}$

4. Solve using Laplace transform, $\frac{dy}{dt} = -2y + 5t$, $y(0) = -4$. Answer: $y = \frac{5t}{2} - \frac{11e^{-2t}}{4} - \frac{5}{4}$.

5. Solve using Laplace transform, $\frac{dy}{dt} + 4y = 3e^{2t}$, $y(0) = -1$. Answer: $y = -\frac{3}{2}e^{-4t} + \frac{e^{2t}}{2}$.

6. Solve using Laplace transform, $\frac{dy}{dt} - 6y = 2e^{2t}$, $y(0) = 2$. Answer: $y = \frac{5e^{6t}}{2} - \frac{e^{2t}}{2}$,

7. Solve using Laplace transform, $\frac{dy}{dt} = 3y + 5e^{3t}$, $y(0) = 2$. Answer: $y = 5e^{3t}t + 2e^{3t}$.

8. Solve using Laplace transform, $\frac{dy}{dt} + y = 2e^{-t}$, $y(0) = 6$. Answer: $y = 2e^{-t}t + 6e^{-t}$.

9. Solve using Laplace transform, $\frac{dy}{dt} - 2y = 2\sin(2t)$, $y(0) = 1$. Answer: $y = \frac{3e^{2t}}{2} - \frac{1}{2}\sin(2t) - \frac{1}{2}\cos(2t)$.

10. Solve using Laplace transform, $\frac{dy}{dt} + 2y = 2\sin(2t) - \cos(2t)$, $y(0) = 0$. Answer: $y = \frac{3e^{-2t}}{4} + \frac{1}{4}\sin(2t) - \frac{3}{4}\cos(2t)$.

11. Solve using Laplace transform, $\frac{dy}{dt} = y + 4t^2$, $y(0) = 3$. Answer: $y = -4t^2 - 8t + 11e^t - 8$.

12. Solve using Laplace transform, $\frac{dy}{dt} = 2y + e^{2t} + 4t^2$, $y(0) = 0$. Answer: $y = -2t^2 + e^{2t}t - 2t + e^{2t} - 1$.

chapter 12

Laplace transform part 2

This chapter deals with solving differential equations that have either the Heaviside function or the Dirac delta function.

12.1   The Heaviside function and its Laplace transform

The Heaviside function is the piecewise function given by

$$u_a(t) = \begin{cases} 1 & \text{if } t \geq a \\ 0 & \text{if } t < a \end{cases}$$

Here, a is a positive real number. Figure 12.1.1 shows the graph of the function $u_6(t)$.

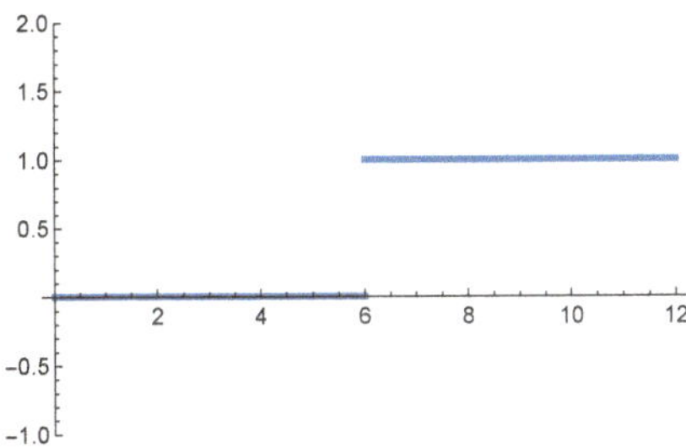

Figure 12.1.1: Graph of the function $u_6(t)$

The computation of the Heaviside function is relatively easy.

Example 110 *The Laplace Transform of the function $u_6(t)$ is given by*

$$\mathcal{L}(u_6(t)) = \int_0^\infty u_6(t)\mathrm{e}^{-st}\,dt = \int_0^6 u_6(t)\mathrm{e}^{-st}\,dt + \int_6^\infty u_6(t)\mathrm{e}^{-st}\,dt = \int_6^\infty \mathrm{e}^{-st}\,dt$$

In the previous computation we have used the fact that the function $u_6(t)$ is zero before $t = 6$ and 1 after $t = 6$. We therefore obtain that

97

$$\mathcal{L}(u_6(t)) = \int_6^\infty e^{-st}\, dt = \left.\frac{e^{-st}}{-s}\right|_{t=6}^{t=\infty} = 0 - \frac{e^{-s6}}{-s} = \frac{e^{-6s}}{s}$$

Following the same reasoning that we did for the case $a = 6$, we have that for any $a > 0$

$$\mathcal{L}(u_a(t)) = \frac{e^{-sa}}{s}$$

12.1.1 u_a shifting of f

When dealing with the Heaviside function, very often we need to get familiar with the following transformation/translation:

Given $a > 0$ we consider the following transformations.

$$f(t) \longrightarrow u_a(t)f(t-a) \quad \text{or} \quad u_a(t)f(t-a) \longrightarrow f(t)$$

We will call the function $f(t)$ the **original function** and the function $u_a(t)f(t-a)$ the u_a-**shifted function**.

Some examples of this transformation are: If $a = 1$ and $f(t) = 6 - t^2$ then the corresponding u_1-shifted function is $u_1(t)(6-(t-1)^2)$. If the shifted function is $u_2(t)\sin(t-2)$, then the original function is $f(t) = \sin(t)$. Figure 12.2.1 shows all four functions

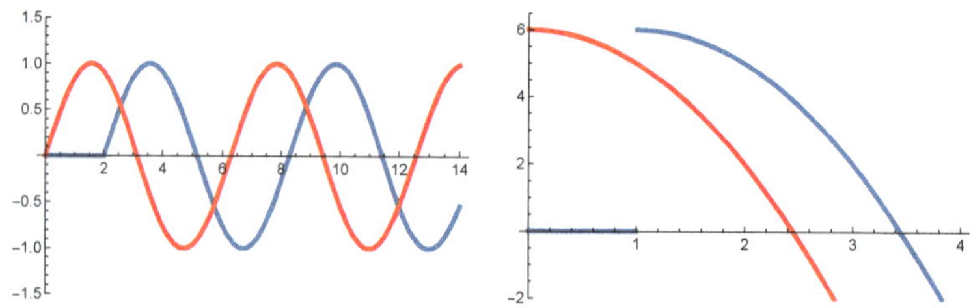

Figure 12.1.2: On the left we show in red the function $f(t) = \sin(t)$ and in blue its u_2 shifting $u_2(t)\sin(t-2)$. On the right we show in red the function $f(t) = 6 - t^2$ and in blue its u_1 shifting $u_1(t)(6 - (t - 1)^2)$.

The following theorem will allows to compute the Laplace transform of u_a shifted functions.

Theorem 17

$$\mathcal{L}(u_a(t)f(t-a)) = e^{-as}\mathcal{L}(f(t))$$

Proof Using the definition we have that

$$\mathcal{L}(u_a(t)f(t-a)) = \int_0^\infty u_a(t)f(t-a)e^{-st}\, dt = \int_a^\infty f(t-a)e^{-st}\, dt$$

By doing the substitution $\tau = t - a$ we obtain that $d\tau = dt$, $\tau = 0$ when $t = a$ and $\tau = \infty$ when $t = \infty$. Therefore

$$\int_a^\infty f(t-a)\mathrm{e}^{-st}\,dt = \int_0^\infty f(\tau)\mathrm{e}^{-s(\tau+a)}\,d\tau = \mathrm{e}^{-sa}\int_0^\infty f(\tau)\mathrm{e}^{-s\tau}\,d\tau = \mathrm{e}^{-sa}\mathcal{L}(f(t))$$

The previous theorem states that in order to compute the Laplace transform of the u_a-shifting of the function $f(t)$, we need to compute the Laplace transform of the original function $f(t)$ and then multiply it by e^{-as}.

Example 111 *Notice that the formula $\mathcal{L}(u_a(t)) = \dfrac{\mathrm{e}^{-as}}{s}$ is a consequence of the previous theorem. Taking $f(t) = 1$ we have that its u_a shifting is $u_a(t)$ and therefore*

$$\mathcal{L}(u_a(t)) = \mathrm{e}^{-as}\mathcal{L}(1) = \mathrm{e}^{-as}\frac{1}{s} = \frac{\mathrm{e}^{-as}}{s}$$

Example 112 *In order to compute $\mathcal{L}(u_2(t)(t-2))$ we notice that the u_2 shifting of $f(t) = t$ is $u_2(t)(t-2)$, therefore,*

$$\mathcal{L}(u_2(t)(t-2))) = \mathrm{e}^{-2s}\mathcal{L}(t) = \mathrm{e}^{-2s}\frac{1}{s^2} = \frac{\mathrm{e}^{-2s}}{s^2}$$

Example 113 *In order to compute $\mathcal{L}(u_4(t)\mathrm{e}^{t-4})$ we notice that the u_4 shifting of $f(t) = \mathrm{e}^t$ is $u_4(t)\mathrm{e}^{t-4}$, therefore*

$$\mathcal{L}(u_4(t)\mathrm{e}^{t-4}) = \mathrm{e}^{-4s}\mathcal{L}(\mathrm{e}^t) = \mathrm{e}^{-4s}\frac{1}{s-1} = \frac{\mathrm{e}^{-4s}}{s-1}$$

Example 114 *In order to compute $\mathcal{L}(6u_3(t)\sin(2(t-3)))$ we notice that the u_3 shifting of $f(t) = 6\sin(2t)$ is $6u_3(t)\sin(2(t-3))$, therefore*

$$\mathcal{L}(6u_3(t)\sin(2(t-3))) = \mathrm{e}^{-3s}\mathcal{L}(6\sin(2t)) = \mathrm{e}^{-3s}\frac{12}{s^2+4} = \frac{12\mathrm{e}^{-3s}}{s^2+4}$$

Now let us solve a differential equation that involves the Heaviside function.

Example 115 *Let us consider the initial value problem $\dfrac{dy}{dt} + 2y = 6u_3(t)\mathrm{e}^{4(t-3)}$ with $y(0) = 1$. In order to solve this problem using the Laplace transform we take the Laplace transform on both sides of the equation and get*

$$s\mathcal{L}(y) - 1 + 2\mathcal{L}(y) = 6\mathrm{e}^{-3s}\frac{1}{s-4}$$

and solving for $\mathcal{L}(y)$ we get,

$$\mathcal{L}(y) = \frac{1}{s+2} + \mathrm{e}^{-3s}\frac{6}{(s+2)(s-4)}$$

Here the advise is to not combine the part that has e^{-as} with the part that does not have it. We now need to compute the Laplace transform inverse of both parts. We have that

$$\frac{6}{(s+2)(s-4)} = \frac{A}{s+2} + \frac{B}{s-4} = \frac{A(s-4) + B(s+2)}{(s+2)(s-4)}$$

A and B must satisfy the equations $A + B = 0$ and $2B - 4A = 6$. The solution is $A = -1$ and $B = 1$. Therefore,

$$\frac{6}{(s+2)(s-4)} = -\frac{1}{s+2} + \frac{1}{s-4} = \mathcal{L}(e^{4t} - e^{-2t})$$

Since the u_3 shifting of $e^{4t} - e^{-2t}$ is the function $u_3(t)\left(e^{4(t-3)} - e^{-2(t-3)}\right)$, then we have that

$$\mathcal{L}(y) = \mathcal{L}(e^{-2t}) + e^{-3s}\mathcal{L}(e^{4t} - e^{-2t}) = \mathcal{L}\left(e^{-2t} + u_3(t)\left(e^{4(t-3)} - e^{-2(t-3)}\right)\right)$$

Therefore $\boxed{y = e^{-2t} + u_3(t)\left(e^{4(t-3)} - e^{-2(t-3)}\right)}$.

Example 116 *To solve the initial value problem $\frac{dy}{dt} - 5y = 4u_2(t)e^{5(t-2)}$ with $y(0) = 9$ we take the Laplace transform on both sides of the equation and we get*

$$s\mathcal{L}(y) - 9 - 5\mathcal{L}(y) = 4e^{-2s}\frac{1}{s-5}$$

and solving for $\mathcal{L}(y)$ we get,

$$\mathcal{L}(y) = \frac{9}{s-5} + e^{-2s}\frac{4}{(s-5)^2} = \mathcal{L}\left(9e^{5t} + 4u_2(t)(t-2)e^{5(t-2)}\right)$$

We have used the fact that $\mathcal{L}(te^{5t}) = \frac{1}{(s-5)^2}$, then $\mathcal{L}\left(u_2(t)(t-2)e^{5(t-2)}\right) = e^{-2s}\frac{1}{(s-5)^2}$.
Therefore $\boxed{y = 9e^{5t} + 4u_2(t)(t-2)e^{5(t-2)}}$.

Example 117 *Let us consider the initial value problem $\frac{dy}{dt} + 3y = 2u_4(t)(2+3t)$ with $y(0) = -6$. Notice that*

$$2 + 3t = 2 + 3(t - 4) + 12 = 14 + 3(t - 4)$$

We have rewritten $2 + 3t$ so we can see that $u_4(t)(2+3t)$ is the u_4 shifting of the function $14 + 3t$. Now taking the Laplace transform on both sides of the equation we get

$$s\mathcal{L}(y) + 6 + 3\mathcal{L}(y) = 2e^{-4s}\left(\frac{14}{s} + \frac{3}{s^2}\right) = e^{-4s}\frac{28s + 6}{s^2}$$

and solving for $\mathcal{L}(y)$ we get,

$$\mathcal{L}(y) = -\frac{6}{s+3} + e^{-4s}\frac{28s + 6}{(s+3)s^2}$$

We have that

$$\frac{28s + 6}{(s+3)s^2} = \frac{A}{s+3} + \frac{B}{s} + \frac{C}{s^2} = \frac{As^2 + Bs(s+3) + C(s+3)}{(s+3)s^2}$$

A, B and C must satisfy the equations $A + B = 0$, $3B + C = 28$ and $3C = 6$. The solution is $A = -\frac{26}{3}$, $B = \frac{26}{3}$ and $C = 2$. Therefore,

$$\frac{28s+6}{(s+3)s^2} = -\frac{\frac{26}{3}}{s+3} + \frac{\frac{26}{3}}{s} + \frac{2}{s^2} = \mathcal{L}\left(-\frac{26}{3}e^{-3t} + \frac{26}{3} + 2t\right)$$

$$
\begin{aligned}
\mathcal{L}(y) &= \mathcal{L}\left(-6e^{-3t}\right) + e^{-4s}\,\mathcal{L}\left(-\frac{26}{3}e^{-3t} + \frac{26}{3} + 2t\right) \\
&= \mathcal{L}\left(-6e^{-3t} + u_4(t)\left(-\frac{26}{3}e^{-3(t-4)} + \frac{26}{3} + 2(t-4)\right)\right)
\end{aligned}
$$

$$\textit{Therefore}\quad \boxed{\, y = -6e^{-3t} + u_4(t)\left(-\frac{26}{3}e^{-3(t-4)} + \frac{2}{3} + 2t\right) \,}$$

Example 118 *Let us consider the initial value problem $\frac{dy}{dt} - y = 4u_2(t)\sin(3(t-2))$ with $y(0) = 7$. Taking the Laplace transform on both sides of the equation we get*

$$s\mathcal{L}(y) - 7 - \mathcal{L}(y) = 4e^{-2s}\frac{3}{s^2+9}$$

and solving for $\mathcal{L}(y)$ we get,

$$\mathcal{L}(y) = \frac{7}{s-1} + e^{-2s}\frac{12}{(s^2+9)(s-1)}$$

We have that

$$\frac{12}{(s^2+9)(s-1)} = \frac{A}{s-1} + \frac{Bs+C}{s^2+9} = \frac{A(s^2+9) + (Bs+C)(s-1)}{(s-1)(s^2+9)}$$

A, B and C must satisfy the equations $A+B = 0$, $C-B = 0$ and $9A - C = 12$. The solution is $A = \frac{6}{5}$, $B = -\frac{6}{5}$ and $C = -\frac{6}{5}$. Therefore,

$$\frac{12}{(s^2+9)(s-1)} = \frac{\frac{6}{5}}{s-1} - \frac{6}{5}\frac{s+1}{s^2+9} = \mathcal{L}\left(\frac{6}{5}e^t - \frac{6}{5}\cos(3t) - \frac{6}{5}\times\frac{1}{3}\sin(3t)\right)$$

$$
\begin{aligned}
\mathcal{L}(y) &= \mathcal{L}\left(7e^t\right) + e^{-2s}\,\mathcal{L}\left(\frac{6}{5}e^t - \frac{6}{5}\cos(3t) - \frac{2}{5}\sin(3t)\right) \\
&= \mathcal{L}\left(7e^t + u_2(t)\left(\frac{6}{5}e^{t-2} - \frac{6}{5}\cos(3(t-2)) - \frac{2}{5}\sin(3(t-2))\right)\right)
\end{aligned}
$$

$$\textit{Therefore}\quad \boxed{\, y = 7e^t + u_2(t)\left(\frac{6}{5}e^{t-2} - \frac{6}{5}\cos(3(t-2)) - \frac{2}{5}\sin(3(t-2))\right) \,}.$$

12.2 The Dirac delta function

Let us start by stating that the Dirac delta function is not a function, it is a *generalized function* or a *distribution*. To understand this notion let us explain how piecewise functions defined on a bounded interval can be viewed as generalized functions. Let us take the piecewise function $f(t) = \begin{cases} t & \text{if } |t| \leq 2 \\ 0 & \text{if } |t| > 2 \end{cases}$. It is clear that $f(t)$ is a function. In order to see $f(t)$ as a generalized function or a distribution we need to see $f(t)$ as a transformation T_f that takes any smooth function g into a number. T_f is defined as

$$T_f(g) = \int_{-\infty}^{\infty} f(t)g(t)\, dt$$

For example, if $g(t) = t$, $T_f(g) = \int_{-\infty}^{\infty} f(t)t\, dt = \int_{-2}^{2} t^2\, dt = \left.\frac{t^3}{3}\right|_{-2}^{2} = \frac{16}{3}$. If $g(t) = \cos(t)$ then $T_f(g) = \int_{-\infty}^{\infty} f(t)\cos(t)\, dt = \int_{-2}^{2} t\cos(t)\, dt = 0$. After taking a look at this example, we define a generalized function as a linear transformation that takes any smooth function into a real number.

> **Definition 18** *For any real number a, the Dirac delta function δ_a is the generalized function that takes the smooth function $g(t)$ into the real number $g(a)$. Most of the time this property is written as*
>
> $$\int_{-\infty}^{\infty} \delta_a(t)g(t)\, dt = g(a)$$

We can see that the sequence of generalized functions coming from the family of piecewise functions $f_{1,a}, f_{2,a} \ldots$ with,

$$f_{n,a}(t) = \begin{cases} n & \text{if } |t - a| \leq \frac{1}{2n} \\ 0 & \text{if } |t - a| > \frac{1}{2n} \end{cases}$$

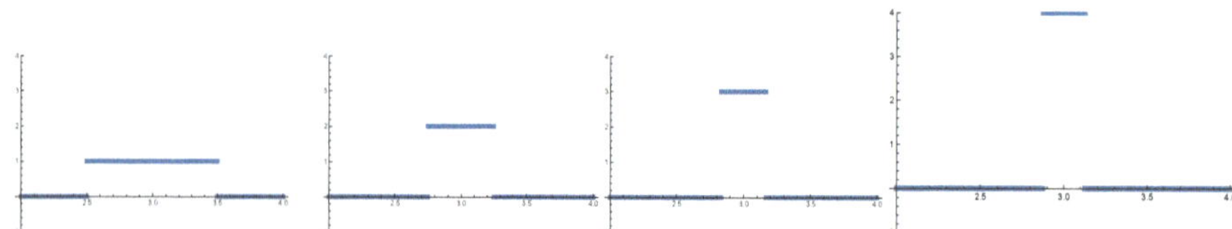

Figure 12.2.1: Graphs of the function $f_{1,3}$, $f_{2,3}$, $f_{3,3}$ and $f_{4,3}$

satisfy that for any smooth function $g(t)$,

$$\lim_{n \to \infty} \int_{-\infty}^{\infty} f_{n,a}(t)g(t)\, dt = \lim_{n \to \infty} \int_{a-\frac{1}{2n}}^{a+\frac{1}{2n}} ng(t)\, dt = g(a) = \int_{-\infty}^{\infty} \delta_a(t)g(t)\, dt$$

Due to the previous limit, the Dirac delta function δ_a is usually viewed as the limit of the functions $f_{n,a}$.

Theorem 18 *The Laplace transform of the Dirac Delta function is given by*

$$\mathcal{L}(\delta_a) = e^{-sa}$$

Proof

$$\mathcal{L}(\delta_a) = \int_0^\infty \delta_a(t)e^{-st}\,dt = e^{-sa}$$

Example 119 *Let us solve the following initial value problem*

$$\frac{dy}{dt} + 2y = 4\delta_3(t) \quad y(0) = 2$$

Taking the Laplace transform on both sides of the equation we get that

$$s\mathcal{L}(y) - 2 + 2\mathcal{L}(y) = 4e^{-3s}$$

Therefore,

$$\mathcal{L}(y) = \frac{2}{s+2} + \frac{4}{s+2}e^{-3s} = \mathcal{L}(2e^{-2t}) + \mathcal{L}(4e^{-2t})e^{-3s} = \mathcal{L}(2e^{-2t} + 4u_3(t)e^{-2(t-3)})$$

and $\boxed{y = 2e^{-2t} + 4u_3(t)e^{-2(t-3)}}$

Example 120 *Let us solve the following initial value problem*

$$\frac{dy}{dt} - 6y = t + 2\delta_3(t) - 3\delta_5(t) \quad y(0) = -1$$

Taking the Laplace transform on both sides of the equation we get that

$$s\mathcal{L}(y) + 1 - 6\mathcal{L}(y) = \frac{1}{s^2} + 2e^{-3s} - 3e^{-5s}$$

Therefore,

$$\mathcal{L}(y) = -\frac{1}{s-6} + \frac{1}{s^2(s-6)} + \frac{2}{s-6}e^{-3s} - \frac{3}{s-6}e^{-5s} = \frac{-s^2+1}{s^2(s-6)} + \frac{2}{s-6}e^{-3s} - \frac{3}{s-6}e^{-5s}$$

But we have that

$$-\frac{1}{s-6} + \frac{1}{s^2(s-6)} = \frac{-s^2+1}{s^2(s-6)} = \frac{A}{s} + \frac{B}{s^2} + \frac{C}{s-6} = \frac{As(s-6) + B(s-6) + Cs^2}{s^2(s-6)}$$

Therefore A, B and C satisfies $A + C = -1$, $B - 6A = 0$ and $-6B = 1$. The solution of the system is $A = -\frac{1}{36}$, $B = -\frac{1}{6}$ and $C = -\frac{35}{36}$. Therefore,

$$\mathcal{L}(y) = \mathcal{L}(-\frac{1}{36} - \frac{t}{6} - \frac{35}{36}e^{6t}) + \mathcal{L}(2e^{6t})e^{-3s} - \mathcal{L}(3e^{6t})e^{-5s}$$

and $\boxed{y = -\frac{1}{36} - \frac{t}{6} - \frac{35}{36}e^{6t} + 2u_3(t)e^{6(t-3)} - 3u_5(t)e^{6(t-5)}}$

12.3 Convolution of two functions

Let us assume that $\mathcal{L}(f(t)) = F(s)$ and $\mathcal{L}(g(t)) = G(s)$. It is natural to wonder what function $h(t)$ satisfies that $\mathcal{L}(h(t)) = F(s)G(s)$. This question can be rephrased as follows: Can we find $\mathcal{L}^{-1}(F(s)G(s))$ in term of $\mathcal{L}^{-1}(F(s))$ and $\mathcal{L}^{-1}(G(s))$. The answer to this question is given by the convolution of two functions. Here is the definition

Definition 19 *Given two functions $f(t)$ and $g(t)$, we define the convolution of f and g as the function*

$$f * g(t) = \int_0^t f(t-u)g(u)du$$

As mentioned at the beginning of this section, the convolution of two functions relates with the Laplace transform.

Theorem 19 *Given two functions $f(t)$ and $g(t)$, we have that $f * g(t) = g * f(t)$ and*

$$\mathcal{L}(f * g(t)) = \mathcal{L}(f(t))\,\mathcal{L}(g(t))$$

Example 121 *If $g(t) = \sin(\omega t)$ and $f(t) = 1$, then*

$$f * g(t) = \int_0^t 1\sin(\omega u)\,du = -\frac{1}{\omega}\cos(\omega u)\Big|_0^t = \frac{1-\cos(\omega t)}{\omega}$$

Notice how Theorem 19 holds true for these two functions. We have that $\mathcal{L}(f * g(t))$ equals to

$$\mathcal{L}(\frac{1-\cos(\omega t)}{\omega}) = \frac{1}{\omega s} - \frac{s}{\omega(s^2+\omega^2)} = \frac{\omega}{s(s^2+\omega^2)} = \frac{1}{s}\frac{\omega}{s^2+\omega^2} = \mathcal{L}(1)\,\mathcal{L}(\sin(\omega t))$$

Example 122 *If $g(t) = e^{3t}$ and $f(t) = t$, then*

$$f * g(t) = \int_0^t (t-u)e^{3u}\,du = \int_0^t te^{3u}\,du - \int_0^t ue^{3u}\,du = \frac{t}{3}e^{3u}\Big|_{u=0}^{u=t} - \int_0^t ue^{3u}\,du$$

Doing integration by parts we have that

$$\int_0^t ue^{3u}\,du = u\frac{e^{3u}}{3}\Big|_{u=0}^{u=t} - \frac{1}{3}\int_0^t e^{3u} = \frac{te^{3t}}{3} - \frac{1}{9}e^{3u}\Big|_{u=0}^{u=t} = \frac{te^{3t}}{3} - \frac{1}{9}e^{3t} + \frac{1}{9}$$

Therefore,

$$f * g(t) = \frac{t}{3}e^{3u}\Big|_{u=0}^{u=t} - \left(\frac{te^{3t}}{3} - \frac{1}{9}e^{3t} + \frac{1}{9}\right) = -\frac{t}{3} + \frac{e^{3t}}{9} - \frac{1}{9}$$

Notice how Theorem 19 holds true for these two functions. We have that $\mathcal{L}(f * g(t))$ equals to

$$\mathcal{L}(-\frac{t}{3} + \frac{e^{3t}}{9} - \frac{1}{9}) = -\frac{1}{3s^2} + \frac{1}{9(s-3)} - \frac{1}{9s} = \frac{1}{s^2(s-3)} = \mathcal{L}(t)\,\mathcal{L}(e^{3t})$$

Example 123 *If $g(t) = e^{2t}$ and $f(t) = 3$, then*

$$f * g(t) = \int_0^t 3e^{2u} \, du = \frac{3}{2}e^{2u} \Big|_0^t = \frac{3}{2}e^{2t} - \frac{3}{2}$$

Notice how Theorem 19 holds true for these two functions. We have that $\mathcal{L}(f * g(t))$ equals to

$$\mathcal{L}(\frac{3}{2}e^{2t} - \frac{3}{2}) = \frac{3}{2(s-2)} - \frac{3}{2s} = \frac{3}{s(s-2)} = \mathcal{L}(3)\,\mathcal{L}(e^{2t})$$

12.4 Homework

1. Solve using Laplace transform, $\frac{dy}{dt} - 4y = 2u_3(t)e^{4(t-3)}$, $y(0) = -2$. Answer: $y = -2e^{4t} + 2u_3(t)e^{4(t-3)}(t-3)$

2. Solve using Laplace transform, $\frac{dy}{dt} - y = u_2(t)3e^{t-2}$, $y(0) = 3$. Answer: $y = 3e^t + 3u_2(t)e^{t-2}(t-2)$.

3. Solve using Laplace transform, $\frac{dy}{dt} + y = 8e^{3(t-7)}u_7(t)$, $y(0) = 3$. Answer: $y = 3e^{-t} + 2u_7(t)\left(e^{3(t-7)} - e^{7-t}\right)$.

4. Solve using Laplace transform, $\frac{dy}{dt} - 6y = 4u_1(t)e^{2(t-1)}$, $y(0) = 2$. Answer: $y = 2e^{6t} - u_1(t)\left(e^{2(t-1)} - e^{6(t-1)}\right)$.

5. Solve using Laplace transform, $\frac{dy}{dt} - 4y = u_4(t)(8t + 3)$, $y(0) = 1$. Answer: $y = e^{4t} + u_4(t)\left(-2t + \frac{37}{4}e^{4(t-4)} - \frac{5}{4}\right)$.

6. Solve using Laplace transform, $\frac{dy}{dt} - 2y = u_1(t)(6t - 3)$, $y(0) = 5$. Answer: $y = 5e^{2t} + u_1(t)\left(3e^{2(t-1)} - 3t\right)$.

7. Solve using Laplace transform, $\frac{dy}{dt} - 3y = u_2(t)(9t + 4)$, $y(0) = 1$. Answer: $y = e^{3t} + u_2(t)\left(-3t + \frac{25}{3}e^{3(t-2)} - \frac{7}{3}\right)$.

8. Solve using Laplace transform, $\frac{dy}{dt} - 3y = 25u_3(t)\sin(4(t - 3))$, $y(0) = -2$. Answer: $y = -2e^{3t} + u_3(t)\left(4e^{3(t-3)} - 3\sin(4(t - 3)) - 4\cos(4(t - 3))\right)$.

9. Solve using Laplace transform, $\frac{dy}{dt} - y = 2 + 3t + 5\delta_3(t) - 7\delta_5(t)$, $y(0) = 5$. Answer: $y = -5 + 10e^t - 3t + 5u_3(t)e^{t-3} - 7u_5(t)e^{t-5}$.

10. Solve using Laplace transform, $\frac{dy}{dt} = 2y + 3e^{4t} + 5\delta_3(t)$, $y(0) = 5$. Answer: $y = \frac{7}{2}e^{2t} + \frac{3}{2}e^{4t} + 5u_3(t)e^{2(t-3)}$.

11. Compute the convolution $h(t) = f * g(t)$ where $f(t) = 1$ and $g(t) = \cos(2t)$. Check that $\mathcal{L}(h(t)) = \mathcal{L}(1)\,\mathcal{L}(\cos(2t))$. Answer $h(t) = \frac{1}{2}\sin(2t)$. We have that $\mathcal{L}(h(t)) = \frac{1}{2}\frac{2}{s^2+4} = \frac{1}{s^2+4}$ which is equal to $\mathcal{L}(1) = \frac{1}{s}$ times $\mathcal{L}(\cos(2t)) = \frac{s}{s^2+4}$

12. Compute the convolution $h(t) = f * g(t)$ where $f(t) = t$ and $g(t) = \sin(3t)$. Check that $\mathcal{L}(h(t)) = \mathcal{L}(f(t))\,\mathcal{L}(g(t))$. Answer $h(t) = \frac{1}{9}(3t - \sin(3t))$. We have that $\mathcal{L}(h(t)) = \frac{1}{9}\left(\frac{3}{s^2} - \frac{3}{s^2+9}\right) = \frac{3}{s^2(s^2+9)}$ which is equal to $\mathcal{L}(t) = \frac{1}{s^2}$ times $\mathcal{L}(\sin(3t)) = \frac{3}{s^2+9}$

chapter 13

Laplace transform part 3

In this section we will be solving second order differential equations using the Laplace transform.

13.1   Solving second order differential equations

Let us start solving a second order differential equation that involves the Heaviside function

Example 124 *Let us consider the initial value problem $\frac{d^2y}{dt^2} + 6\frac{dy}{dt} + 5y = 8u_3(t)e^{-3(t-3)}$ with $y(0) = -2$ and $y'(0) = 1$. Taking the Laplace transform on both sides of the equation we get*

$$s^2\mathcal{L}(y) + 2s - 1 + 6(s\mathcal{L}(y) + 2) + 5\mathcal{L}(y) = 8e^{-3s}\frac{1}{s+3}$$

and solving for $\mathcal{L}(y)$ we get,

$$\mathcal{L}(y) = \frac{-2s - 11}{(s^2 + 6s + 5)} + e^{-3s}\frac{8}{(s+3)(s^2 + 6s + 5)}$$

Using the fact that $s^2 + 6s + 5 = (s+1)(s+5)$ we obtain that

$$\frac{-2s - 11}{(s^2 + 6s + 5)} = \frac{-2s - 11}{(s+1)(s+5)} = \frac{A}{s+1} + \frac{B}{s+5} = \frac{A(s+5) + B(s+1)}{(s+1)(s+5)}$$

A and B must satisfy the equations $A+B = -2$ and $5A+B = -11$. The solution is $A = -9/4$ and $B = 1/4$. Therefore,

$$\frac{-2s - 11}{(s^2 + 6s + 5)} = -\frac{9/4}{s+1} + \frac{1/4}{s+5} = \mathcal{L}(\frac{1}{4}e^{-5t} - \frac{9}{4}e^{-t})$$

On the other hand we have that

$$\begin{aligned}
\frac{8}{(s+3)(s^2 + 6s + 5)} &= \frac{A}{s+1} + \frac{B}{s+5} + \frac{C}{s+3} \\
&= \frac{A(s+5)(s+3) + B(s+1)(s+3) + C(s+1)(s+5)}{(s+3)(s+1)(s+5)}
\end{aligned}$$

A, B and C must satisfy $A + B + C = 0$, $8A + 4B + 6C = 0$ and $15A + 3B + 5C = 8$. The solution is $A = 1$, $B = 1$ and $C = -2$. Therefore,

$$\frac{8}{(s+3)(s^2+6s+5)} = \frac{1}{s+1} + \frac{1}{s+5} - \frac{2}{s+3} = \mathcal{L}\left(e^{-t} + e^{-5t} - 2e^{-3t}\right)$$

and

$$e^{-3s}\frac{8}{(s+3)(s^2+6s+5)} = \mathcal{L}\left(u_3(t)\left(e^{-(t-3)} + e^{-5(t-3)} - 2e^{-3(t-3)}\right)\right)$$

$$\boxed{\textit{Therefore}\ \ y = \frac{1}{4}e^{-5t} - \frac{9}{4}e^{-t} + u_3(t)\left(e^{-(t-3)} + e^{-5(t-3)} - 2e^{-3(t-3)}\right)}$$

The following theorem will allow us to solve a bigger variety of differential equations.

Theorem 20 *If $\mathcal{L}(f(t)) = F(s)$, then $\mathcal{L}(e^{at}f(t)) = F(s-a)$.*

Proof We have that

$$\mathcal{L}(e^{at}f(t)) = \int_{t=0}^{t=\infty} e^{at}f(t)e^{-st}\,dt = \int_{t=0}^{t=\infty} f(t)e^{at-st}\,dt = \int_{t=0}^{t=\infty} f(t)e^{-(s-a)t}\,dt = F(s-a)$$

As a consequence of the previous theorem we have the following,

Example 125 *Since $\mathcal{L}(\cos(\omega t)) = \frac{s}{s^2+\omega^2}$ then we have that*

$$\mathcal{L}\left(e^{at}\cos(\omega t)\right) = \frac{s-a}{(s-a)^2+\omega^2}$$

Since $\mathcal{L}(\sin(\omega t)) = \frac{\omega}{s^2+\omega^2}$ then we have that

$$\mathcal{L}\left(e^{at}\sin(\omega t)\right) = \frac{\omega}{(s-a)^2+\omega^2}$$

Since $\mathcal{L}(t^n) = \frac{n!}{s^{n+1}}$ then we have that

$$\mathcal{L}\left(e^{at}t^n\right) = \frac{n!}{(s-a)^{n+1}}$$

Let us practice some Laplace transform inverse using the previous formulas. A technique that we will be required to do the following computations is the one of completing squares.

Example 126 *In order to compute $\mathcal{L}^{-1}\left(\frac{3s-14}{s^2+4s+13}\right)$ we first point out that*

$$s^2 + 4s + 13 = s^2 + 4s + 4 + 9 = (s+2)^2 + 9$$

Therefore,

$$\frac{3s-14}{s^2+4s+13} = \frac{3(s+2)-6-14}{(s+2)^2+9} = \frac{3(s+2)-20}{(s+2)^2+9} = \frac{3(s+2)}{(s+2)^2+9} - \frac{20}{(s+2)^2+9}$$

$$= \frac{3(s+2)}{(s+2)^2+9} - \frac{20}{3}\frac{3}{(s+2)^2+9} = \mathcal{L}\left(3e^{-2t}\cos(3t) - \frac{20}{3}e^{-2t}\sin(3t)\right)$$

Example 127 *In order to compute $\mathcal{L}^{-1}\left(\frac{-3s+10}{s^2+6s+14}\right)$ we first point out that*

$$s^2 + 6s + 14 = s^2 + 6s + 9 + 5 = (s+3)^2 + 5$$

Therefore,

$$
\begin{aligned}
\frac{-3s+10}{s^2+6s+14} &= \frac{-3s+10}{(s+3)^2+5} = \frac{-3(s+3)+9+10}{(s+3)^2+5} = \frac{-3(s+3)}{(s+3)^2+5} + \frac{19}{(s+3)^2+5} \\
&= \mathcal{L}\left(-3e^{-3t}\cos(\sqrt{5}t) + \frac{19}{\sqrt{5}}e^{-3t}\sin(\sqrt{5}t)\right)
\end{aligned}
$$

Let us now solve a differential equation that uses arguments similar to the previous two examples.

Example 128 *Let us consider the initial value problem $\frac{d^2y}{dt^2} - 4\frac{dy}{dt} + 20y = 64u_7(t)\sin 2(t-7)$ with $y(0) = 2$ and $y'(0) = 3$. Taking the Laplace transform on both sides of the equation and we get*

$$s^2\mathcal{L}(y) - 2s - 3 - 4(s\mathcal{L}(y) - 2) + 20\mathcal{L}(y) = e^{-7s}\frac{128}{s^2+4}$$

Solving for $\mathcal{L}(y)$ we obtain

$$\mathcal{L}(y) = \frac{2s-5}{s^2-4s+20} + e^{-7s}\frac{128}{(s^2+4)(s^2-4s+20)}$$

We have that

$$
\begin{aligned}
\frac{2s-5}{s^2-4s+20} &= \frac{2(s-2)-1}{(s-2)^2+16} = 2\frac{s-2}{(s-2)^2+16} - \frac{1}{4}\frac{4}{(s-2)^2+16} \\
&= \mathcal{L}\left(e^{2t}\left(2\cos(4t) - \frac{1}{4}\sin(4t)\right)\right)
\end{aligned}
$$

On the other hand, we have that

$$
\begin{aligned}
\frac{128}{(s^2+4)(s^2-4s+20)} &= \frac{As+B}{s^2-4s+20} + \frac{Cs+D}{s^2+4} \\
&= \frac{(As+B)(s^2+4) + (Cs+D)(s^2-4s+20)}{(s^2+4)(s^2-4s+20)}
\end{aligned}
$$

A direct computation shows that A, B, C and D must satisfy the following system of equations

$$
\begin{cases}
A + C & = 0 \\
B + D - 4C & = 0 \\
4A + 20C - 4D = 0 \\
4B + 20D = 128
\end{cases}
$$

The solution of the system above is $A = -\frac{8}{5}$, $B = 0$, $C = \frac{8}{5}$ and $D = \frac{32}{5}$. Therefore,

$$\frac{128}{(s^2+4)(s^2-4s+20)} = \frac{-\frac{8}{5}s}{s^2-4s+20} + \frac{\frac{8}{5}s+\frac{32}{5}}{s^2+4}$$

$$= \frac{-\frac{8}{5}(s-2)-\frac{16}{5}}{(s-2)^2+16} + \frac{8}{5}\frac{s}{s^2+4} + \frac{32}{5}\frac{1}{s^2+4}$$

$$= -\frac{8}{5}\frac{(s-2)}{(s-2)^2+16} - \frac{4}{5}\frac{4}{(s-2)^2+16} + \frac{8}{5}\frac{s}{s^2+4} + \frac{16}{5}\frac{2}{s^2+4}$$

$$= \mathcal{L}\left(e^{2t}\left(-\frac{8}{5}\cos(4t)-\frac{4}{5}\sin(4t)\right) + \frac{8}{5}\cos(2t) + \frac{16}{5}\sin(2t)\right)$$

Therefore

$$\mathcal{L}(y) = \mathcal{L}(e^{2t}(2\cos(4t)-\frac{1}{4}\sin(4t))) +$$

$$e^{-7s}\mathcal{L}\left(e^{2t}\left(-\frac{8}{5}\cos(4t)-\frac{4}{5}\sin(4t)\right) + \frac{8}{5}\cos(2t) + \frac{16}{5}\sin(2t)\right)$$

and

$$y = e^{2t}(2\cos(4t)-\frac{1}{4}\sin(4t)) +$$

$$u_7(t)\left(e^{2(t-7)}\left(-\frac{8}{5}\cos(4(t-7))-\frac{4}{5}\sin(4(t-7))\right) + \frac{8}{5}\cos(2(t-7)) + \frac{16}{5}\sin(2(t-7))\right)$$

Example 129 *Let us consider the initial value problem $\frac{d^2y}{dt^2} + 4\frac{dy}{dt} + 20y = 41u_3(t)e^{3(t-3)}$ with $y(0)=2$ and $y'(0)=-4$. Taking the Laplace transform on both sides of the equation we get*

$$s^2\mathcal{L}(y) - 2s + 4 + 4(s\mathcal{L}(y)-2) + 20\mathcal{L}(y) = e^{-3s}\frac{41}{s-3}$$

Solving for $\mathcal{L}(y)$ we obtain

$$\mathcal{L}(y) = \frac{2s+4}{s^2+4s+20} + e^{-3s}\frac{41}{(s-3)(s^2+4s+20)}$$

We have that

$$\frac{2s+4}{s^2+4s+20} = \frac{2(s+2)}{(s+2)^2+16} = 2\frac{s+2}{(s+2)^2+16} = \mathcal{L}\left(2e^{-2t}\cos(4t)\right)$$

On the other hand, we have that

$$\frac{41}{(s-3)(s^2+4s+20)} = \frac{As+B}{s^2+4s+20} + \frac{C}{s-3} = \frac{(As+B)(s-3)+C(s^2+4s+20)}{(s-3)(s^2+4s+20)}$$

A direct computation shows that A, B and C must satisfy the following system of equations

$$\begin{cases} A + C & = 0 \\ B - 3A + 4C & = 0 \\ -3B + 20C = 41 \end{cases}$$

The solution of the system above is $A = -1$, $B = -7$ and $C = 1$. Therefore,

$$\begin{aligned}
\frac{41}{(s-3)(s^2+4s+20)} &= \frac{-s-7}{s^2+4s+20} + \frac{1}{s-3} \\
&= \frac{-(s+2)+2-7}{(s+2)^2+16} + \frac{1}{s-3} \\
&= -\frac{(s+2)}{(s+2)^2+16} - \frac{5}{4}\frac{4}{(s+2)^2+16} + \frac{1}{s-3} \\
&= \mathcal{L}\left(e^{-2t}\left(-\cos(4t) - \frac{5}{4}\sin(4t) \right) + e^{3t} \right)
\end{aligned}$$

Therefore,

$$\mathcal{L}(y) = \mathcal{L}(2e^{-2t}\cos(4t)) + e^{-3s}\left(e^{3t} - e^{-2t}\left(\cos(4t) + \frac{5}{4}\sin(4t) \right) \right)$$

and

$$\boxed{y = 2e^{-2t}\cos(4t)) + u_3(t)\left(e^{3(t-3)} - e^{-2(t-3)}\left(\cos(4(t-3)) + \frac{5}{4}\sin(4(t-3)) \right) \right)}$$

13.2 Homework

1. Solve $\frac{d^2y}{dt^2} - 5\frac{dy}{dt} + 4y = u_1(t)6e^{2(t-1)}$, $y(0) = -2$, $y'(0) = 1$. Answer:

$$y = -3e^t + e^{4t} + u_1(t)\left(2e^{t-1} - 3e^{2(t-1)} + e^{4(t-1)} \right)$$

2. Solve $\frac{d^2y}{dt^2} - \frac{dy}{dt} - 6y = u_2(t)30e^{4(t-2)}$, $y(0) = 2$, $y'(0) = -2$. Answer:

$$y = \frac{8e^{-2t}}{5} + \frac{2e^{3t}}{5} + u_2(t)\left(e^{-2(t-2)} - 6e^{3(t-2)} + 5e^{4(t-2)} \right)$$

3. Solve $\frac{d^2y}{dt^2} - 2\frac{dy}{dt} + 5y = u_1(t)13e^{-2(t-1)}$, $y(0) = 1$, $y'(0) = 3$. Answer:

$$y = e^t\sin(2t) + e^t\cos(2t) + u_1(t)\left(e^{-2(t-1)} + \tfrac{3}{2}e^{t-1}\sin(2(t-1)) - e^{t-1}\cos(2(t-1)) \right)$$

4. Solve $\frac{d^2y}{dt^2} + 2\frac{dy}{dt} + 5y = 20u_4(t)\sin(t-4)$, $y(0) = 1$, $y'(0) = -3$. Answer:

$$y = -e^{-t}\sin(2t) + e^{-t}\cos(2t) +$$
$$u_4(t)\left(4\sin(t-4) - e^{-(t-4)}\sin(2(t-4)) - 2\cos(t-4) + 2e^{-(t-4)}\cos(2(t-4)) \right)$$

5. Solve $\frac{d^2y}{dt^2} + 8\frac{dy}{dt} + 17y = 425u_2(t)\sin(2(t-2))$, $y(0) = 6$, $y'(0) = 4$. Answer:

$$y = 28e^{-4t}\sin(t) + 6e^{-4t}\cos(t) +$$
$$u_2(t)\left(38e^{-4(t-2)}\sin(t-2) + 13\sin(2(t-2)) + 16e^{-4(t-2)}\cos(t-2) - 16\cos(2(t-2))\right)$$

6. Solve $\frac{d^2y}{dt^2} + 6\frac{dy}{dt} + 10y = 65u_8(t)e^{5(t-8)}$, $y(0) = 3$, $y'(0) = -4$. Answer:

$$y = 5e^{-3t}\sin(t) + 3e^{-3t}\cos(t) + u_8(t)\left(e^{5(t-8)} - 8e^{-3(t-8)}\sin(t-8) - e^{-3(t-8)}\cos(t-8)\right)$$